Geomorphological Processes and Landscape Change

RGS-IBG Book Series

The *Royal Geographical Society (with the Institute of British Geographers) Book Series* provides a forum for scholarly monographs and edited collections of academic papers at the leading edge of research in human and physical geography. The volumes are intended to make significant contributions to the field in which they lie, and to be written in a manner accessible to the wider community of academic geographers. Some volumes will disseminate current geographical research reported at conferences or sessions convened by Research Groups of the Society. Some will be edited or authored by scholars from beyond the UK. All are designed to have an international readership and to both reflect and stimulate the best current research within geography.

The books will stand out in terms of:

- the quality of research
- their contribution to their research field
- their likelihood to stimulate other research
- being scholarly but accessible

Published

Geomorphological Processes and Landscape Change: Britain in the Last 1000 Years
David L. Higgitt and E. Mark Lee (eds)

Globalizing South China
Carolyn L. Cartier

Geomorphological Processes and Landscape Change: Britain in the Last 1000 Years

Edited by

David L. Higgitt
and
E. Mark Lee

BLACKWELL
Publishers

Copyright © Blackwell Publishers Ltd 2001
Editorial matter and arrangement copyright © David L. Higgitt and E. Mark Lee 2001

The moral right of David L. Higgitt and E. Mark Lee to be identified as authors of the editorial material has been asserted in accordance with the Copyright, Designs and Patents Act 1988.

First published 2001

2 4 6 8 10 9 7 5 3 1

Blackwell Publishers Ltd
108 Cowley Road
Oxford OX4 1JF
UK

Blackwell Publishers Inc.
350 Main Street
Malden, Massachusetts 02148
USA

British Library Cataloguing in Publication Data
A CIP catalogue record for this book is available from the British Library.

Library of Congress Cataloging-in-Publication Data
Geomorphological processes and landscape change : Britain in the last 1000 years / edited by David L. Higgitt and E. Mark Lee.
    p. cm.
Includes bibliographical references (p.   )
    ISBN 0-631-22273-1 (hbk : alk. paper)
    1. Geomorphology—Great Britain. I. Higgitt, David L. II. Lee, E. Mark. III. Title.
    GB436.G7 G45 2001
    551.41′0941—dc21

                                                                2001001196

Typeset in 10.5 on 12 pt Plantin
by SetSystems Ltd, Saffron Walden, Essex

This book is printed on acid-free paper.

Contents

Figures

Tables

Contributors

Denys Brunsden, *Emeritus*, Department of Geography, Kings College London, The Strand, London, WC2R 2LS, UK

Martin G. Evans, School of Geography, University of Manchester, Oxford Road, Manchester, M13 9PL, UK

Ian D. L. Foster, Centre for Environmental Research and Consultancy, NES (Geography), Coventry University, Priory Street, Coventry, CV1 5FB, UK

David L. Higgitt, Department of Geography, University of Durham, Science Laboratories, South Road, Durham, DH1 3LE, UK

Janet M. Hooke, Department of Geography, University of Portsmouth, Buckingham Building, Lion Terrace, Portsmouth, PO1 3HE, UK

David K. C. Jones, Department of Geography and Environment, The London School of Economics and Political Science, Houghton Street, London, WC2A 2AE, UK

E. Mark Lee, Department of Marine Sciences and Coastal Management, University of Newcastle upon Tyne, Ridley Building, Newcastle upon Tyne, NE1 7RU, UK

Barbara T. Rumsby, Department of Geography, The University of Hull, Cottingham Road, Hull, HU6 7RU, UK

Jeff Warburton, Department of Geography, University of Durham, Science Laboratories, South Road, Durham, DH1 3LE, UK

Geomorphological Processes and Landscape Change

RGS-IBG Book Series

The *Royal Geographical Society (with the Institute of British Geographers) Book Series* provides a forum for scholarly monographs and edited collections of academic papers at the leading edge of research in human and physical geography. The volumes are intended to make significant contributions to the field in which they lie, and to be written in a manner accessible to the wider community of academic geographers. Some volumes will disseminate current geographical research reported at conferences or sessions convened by Research Groups of the Society. Some will be edited or authored by scholars from beyond the UK. All are designed to have an international readership and to both reflect and stimulate the best current research within geography.

The books will stand out in terms of:

- the quality of research
- their contribution to their research field
- their likelihood to stimulate other research
- being scholarly but accessible

Published

Geomorphological Processes and Landscape Change: Britain in the Last 1000 Years
David L. Higgitt and E. Mark Lee (eds)

Globalizing South China
Carolyn L. Cartier

Geomorphological Processes and Landscape Change: Britain in the Last 1000 Years

Edited by

David L. Higgitt

and

E. Mark Lee

BLACKWELL
Publishers

Copyright © Blackwell Publishers Ltd 2001
Editorial matter and arrangement copyright © David L. Higgitt and E. Mark Lee 2001

The moral right of David L. Higgitt and E. Mark Lee to be identified as authors of the
editorial material has been asserted in accordance with the Copyright, Designs and
Patents Act 1988.

First published 2001

2 4 6 8 10 9 7 5 3 1

Blackwell Publishers Ltd
108 Cowley Road
Oxford OX4 1JF
UK

Blackwell Publishers Inc.
350 Main Street
Malden, Massachusetts 02148
USA

British Library Cataloguing in Publication Data
A CIP catalogue record for this book is available from the
British Library.

Library of Congress Cataloging-in-Publication Data
Geomorphological processes and landscape change : Britain in the last
1000 years / edited by David L. Higgitt and E. Mark Lee.
    p. cm.
Includes bibliographical references (p.   )
  ISBN 0-631-22273-1 (hbk : alk. paper)
  1. Geomorphology—Great Britain. I. Higgitt, David L. II. Lee, E.
Mark. III. Title.
  GB436.G7 G45 2001
  551.41′0941—dc21

                                           2001001196

Typeset in 10.5 on 12 pt Plantin
by SetSystems Ltd, Saffron Walden, Essex
Printed in Great Britain by TJ International, Padstow, Cornwall

This book is printed on acid-free paper.

Figure Acknowledgements

The editors and publishers are grateful to the following for permission to reproduce copyright material.

Figure 2.1(b) is reproduced from J. Sheenan 1989: Holocene crustal movements and sea-level changes in Great Britain. *Journal of Quaternary Science*, 4, 77–89, by permission of Pearson Education Limited. Figure 2.2(a) is reproduced from J. T. Greensmith and E. V. Tucker 1973: Holocene transgressions and regressions on the Essex coast, outer Thames estuary. *Geologie en Mijnbouw*, 52, 193–202, figure 1, copyright © 1973 Kluwer Academic Publishers, with kind permission from Kluwer Academic Publishers. Figure 2.2(b) is reproduced from R. W. G. Carter 1988: *Coastal Environments: an Introduction to the Physical, Ecological and Cultural Systems of Coastlines*, by permission of Academic Press Limited, London. Figure 2.5 is reproduced from H. H. Lamb 1997: *Climate, History and the Modern World*, by permission of Routledge. Figure 2.6 is reproduced from B. W. Cunliffe 1980: The evolution of Romney Marsh: a preliminary statement. In F. H. Thompson (ed.), *Archaeology and Coastal Change*, 37–55, by permission of the Society of Antiquaries, London. Figure 2.8(a) is reproduced from P. Doody and B. Barnett (eds) 1987: *The Wash and its Environment*. Peterborough: Nature Conservancy Council, by permission of English Nature. Figure 2.8(b) is reproduced from P. Burrin 1985: Holocene alluviation in southeast England and some implications for palaeohydrological studies. *Earth Surface Processes and Landforms*, 10, 252–72, copyright © 1985 John Wiley & Sons Limited, by permission of John Wiley & Sons Limited. Figure 4.2 is reproduced from A. G. Brown and T. A. Quine 1999: Fluvial processes and environmental change: an overview. In A. G. Brown and T. A. Quine (eds), *Fluvial Processes and Environmental Change*, 1–27, copyright © 1999 John Wiley & Sons Limited, by permission of John Wiley & Sons Limited. Figure 5.1 is reproduced from H. C. Darby and E. M. J. Campbell 1962: *The Domesday Geography of South-East England*, by permission of Cambridge University Press. Figures 6.1(b) and 6.1(c) are reproduced from J. Pethick 1992: Natural change. In M. G. Barrett (ed.), on behalf of the Institution of Civil Engineers, *Coastal Zone Planning and Management*. London: Thomas Telford, 49–63, by permission of the Institution of Civil Engineers. Figure 6.2 is reproduced from J. Pethick 1996: Coastal slope development: temporal and spatial periodicity in the Holderness Cliff Recession. In M. G. Anderson and S. M. Brooks (eds), *Advances in Hillslope Processes*, Vol. 2, 897–917, copyright © 1996 John Wiley & Sons Limited, by permission of John Wiley & Sons Limited. Figures 6.3 and 6.4 are reproduced from J. Pethick 1996: The geomorphology of mudflats. In K. F. Nordstrom and C. T. Roman (eds), *Estuarine Shores: Evolution, Environments and Human Alterations*, copyright © 1996 John Wiley & Sons Limited, by permission of John Wiley & Sons Limited. Figure 6.5 is reproduced from R. W. G. Carter 1992: Coastal conservation. In M. G. Barrett (ed.), on behalf of the Institution of Civil Engineers, *Coastal Zone Planning and Management*. London: Thomas Telford, 21–48, by permission of the Institution of Civil Engineers. Figure 8.2 is reproduced from D. Ashbridge 1995: Processes of river bank erosion and their contribution to the suspended sediment load of the River Culm, Devon. In I. D. L. Foster, A. M. Gurnell and B. W. Webb (eds), *Sediment and Water Quality in River Catchments*, 229–45, copyright © 1995 John Wiley & Sons Limited, by permission of John Wiley & Sons Limited.

Chapter 1

A Brief Time of History

David L. Higgitt

1.1 Aims and Objectives

The landscape of Britain displays an enormous variety of scenery in a comparatively small space. The diversity of rock types, the influence of geological structure and the impact of successive glaciations in moulding the uplands and redistributing material provide the student of geomorphology with a wide assortment of landforms to inspect and appreciate. Numerous volumes have been conceived to celebrate and explain the geology and landforms of Britain (Trueman, 1949; Goudie and Gardner, 1992; Goudie and Brunsden, 1994). It might be considered somewhat ironic that so many geomorphologists should inhabit an island where the inland contemporary geomorphology is, to all intents, dominated by low-intensity processes operating on an essentially relict landscape. Tectonic processes are almost negligible, with volcanic activity last experienced in the Tertiary. The impact of natural hazards is mercifully low, if not entirely absent. In short, there are more exciting parts of the Earth to investigate active geomorphological processes. What relevance, therefore, does a volume on geomorphological processes in Britain during the last 1000 years hold?

Students of scenery, in its broadest sense, can point to the varied and profound changes that have occurred during the last millennium. Regional forest clearance may have been well under way by AD 1000, but deforestation continued, and agricultural activity expanded into the hills as climate warmed, then switched to grazing as the climate cooled. Many villages and towns expanded, while some settlements were abandoned. Fields were enclosed. Mining produced both wealth and large quantities of sediment supply. Water resources have been exploited from the development of mills through to the construction of reservoirs. Again, there are numerous volumes that examine the evolution of agriculture, townscapes and settlement patterns and the shaping of the cultural landscape of Britain (Hoskins, 1955; Coones and Patten, 1986; Whyte and Whyte, 1991). In these narratives the physical landscape is often treated as a backcloth upon which human activity adapted and created scenery. But to what extent did the operation of geomorphological processes in the physical landscape constrain human use of the land or react to human-induced changes? What evidence is available to reconstruct geomorphological process activity in the past 1000 years?

The principal aim of the volume is to provide an overview of the nature of geomorphological process activity and landscape change in Britain over the last 1000 years. The impact of the driving forces of climate change and human action are considered through the framework of the sediment cascade. The production of sediments through weathering processes generates a supply that can be transferred from slopes to channels and transported through river basins to the coast. The relationship between the production and availability of sediment and the ability of prevailing geomorphological processes to transfer that sediment is examined. The contents are organized by considering process components of the sediment cascade, starting with hillslopes and moving down-basin towards the coast. In evaluating the evidence for sediment transfer, it becomes clear that the period of the last 1000 years offers a number of challenges to geomorphologists, both in terms of the techniques and methodologies available for identifying and reconstructing data and in terms of the conceptual approaches to evaluating change over this timescale. The book, therefore, aims to examine the nature of the evidence for geomorphological change, to evaluate the role of human impact and of climate control, and to assess the suitability of available techniques and concepts to interpret environmental change. Although the focus is geographically constrained on the landscape of Britain, the approaches to evaluating millennial-scale changes have much wider application. The relatively intensely studied field sites in Britain provide some experience of the information about past geomorphological processes that can be yielded but also identify the considerable gaps in our present knowledge. Such information is not only of interest to geomorphologists but, increasingly, has a role in complementing the interpretations of archaeologists, historians and ecologists. Understanding how landscape components respond to forcing conditions of land use change, to the climatic regime and to individual events has implications for evaluating the hazards that geomorphological processes might pose for society and for the conservation and management of habitat and scenery. A growing collaboration between geomorphologists, ecologists and engineers has emerged in recent years. This leads to the final objective of the book: to evaluate the role of geomorphology for the mitigation of natural hazards and for the promotion of environmental management.

1.2   The Millennium in Perspective

A rationalist might reasonably ask whether the theme presented within this volume is merely a reaction to the millennium bandwagon. Separating the most recent span of 1000 years risks compartmentalizing geomorphological change that has been progressing throughout the Holocene. It is, of course, absurd to suppose that any socially constructed span of time has any pre-ordained or unique relevance to the operation of geomorphological processes and the characteristics of landscape. There are, however, several reasons why it is worth reflecting on the last 1000 years at this juncture, among which is the empathy for reflecting about the past and speculating about the future that has been displayed in many quarters as the 'millennial theme'. Five features of note are the significance of human impact, the need to integrate understanding of contemporary process dynamics with reconstructed environments, the growing appreciation of historical geomorphology for conservation purposes, the challenge of managing risk from natural hazards and the implications of past and future climate change.

First, the span of the last 1000 years has witnessed significant human impact on the landscape. This introduction has been written in a university a few hundred metres from the magnificent cathedral of Durham. Now designated a World Heritage Site, building of the present structure began in 1093. This enduring symbol of architectural achievement has stood for almost all of the last millennium – the period over which geomorphological change is to be considered. The imposing splendour of Durham's 'massive piles, half Church of God, half castle 'gainst the Scot' (as immortalized in the poem by Sir Walter Scott), owes much to its location high above the incised meander of the River Wear. The meander loop is cut through the Coal Measures, most probably as a result of meltwater diversion during the Late Glacial. The combined vista of cathedral, steep wooded gorge and river projects an image of permanence and durability that has repelled change for ages past – a massive display of Norman strength (Pocock, 1996) defended by the steep slopes of the peninsula. There is little doubt, geomorphologically, that the physiography of the River Wear meander gorge owes much to the combination of meltwater discharge and isostatic recovery. The Wear has a buried rock-cut channel, formed during an earlier period of lower sea level, which was choked by glacial diamict, forcing the river to adopt a new course when the ice wasted away. Subsequent isostatic uplift incised the meander loop and it has undergone comparatively minimal topographic change throughout the Holocene. Like much of the British landscape, the landforms are fundamentally inherited from formative processes that operated in the Late Glacial. The last 1000 years might, at first glance, appear irrelevant to the scene. It can be argued that the significance of geomorphological processes may have been limited since the foundation stones of the cathedral were laid in the eleventh century, but that there has been a range of process activity experienced during this period.

The walls of the cathedral, fashioned primarily from local sandstone, exhibit some intricate patterns of weathering. It would be possible, with care, to estimate the net rate of disintegration, as has been recently attempted for fifteenth- and sixteenth-century structures on the south-west coast of England (Mottershead, 2000) which probably amount to a few millimetres per century. Close to the western façade of the cathedral, a rotational failure measuring about 10 m across by 20 m in length closed a hillside footpath in 1999. It is apparent that small-scale mass movements have affected several of the slopes of the peninsula, causing some inconvenience to the foundations of the buildings on the peninsula, but their significance is tiny compared to the impact of quarrying and coal mining. Neither is there clear evidence for any recent change in the course of the Wear, which is effectively locked into its postglacial gorge. The history of the bridges that cross the Wear around its meander loop (Elvet, Prebends and Framwellgate) says something about fluvial processes. Remnants of piers of the original twelfth-century Elvet Bridge suggest a possible small lateral shift of the channel as it enters the gorge, but it is clear that the millennium has seen little physical change to the view. Of more significance is the geomorphological work accomplished through the transport of water and sediment, and the action of humans to harness and control it. Archival records of ancient floods are scarce, but in the North-East of England much information has been assembled by Archer (1992). As early as AD 1400, Framwellgate Bridge had been swept away in a flood. Documentary records indicate a number of floods in the seventeenth century, including two major events in the 1680s – one of the wettest decades in the reconstructed millennial climate record. Upstream of the gorge, where the river is incised into a wide floodplain, the bridge at Shincliffe was destroyed in February 1753, while Sunderland Bridge was damaged three times in the early eighteenth century (Archer, 1992). The most remarkable flood, however, was that of 17 November 1771, by far the largest on record, which also affected the neighbouring Tyne and Tees. Several bridges in Weardale were destroyed, while in Durham itself, the medieval Elvet Bridge was severely damaged and Prebends Bridge completely swept away (Archer, 1992). A succession of flood events followed in the 1820s and many others have followed in the past two centuries. The year AD 2000 has witnessed two major flood events on the Wear, on 4 June and 6–7 November. Geomorphologically, each flood resulted in substantial amounts of bank collapse upstream of Durham, although the evidence from the summer flood was quickly masked by vegetation. Academically, the June flood had a dramatic impact on a riparian venue that was being used for university examinations! While the summer flood was reasonably localized, the November flood was but one of many that affected many parts of Britain. From the millennium perspective, it can be noted that floods on this particular English river have occurred at different times of year, and have resulted from rapid snowmelt, cyclonic rainfall or convectional storms. Although the physical change to the landscape view of the cathedral during its lifetime is limited, fluvial processes have conveyed substantial amounts of sediment past the scene, especially in the eighteenth and nineteenth centuries when metal mining augmented sediment supply, and floods have caused considerable damage to riverside property and their inhabitants. Deliberate modification is also apparent. The original flood embankments upstream of Durham are of Cistercian origin. There are two ancient weirs on the meander loop, providing intake for mills. The lower weir (Framwellgate) was raised by over 1 m in 1935. Downstream of this weir, the channel was widened in 1964 and dredged in 1967 (Archer, 1992). The interweaving of deliberate and incidental human interaction with episodic natural processes is a pervading theme for geomorphology at the 1000 year timescale.

Second, the timescale of 1000 years provides an opportunity to address the apparent schism that has developed between process-orientated geomorphological studies and stratigraphical analysis of Late Quaternary environmental change. In some locations much more is known about the nature of the environment through the Late Glacial and early Holocene than during the last millennium. The development and refinement of dating techniques and modelling approaches offer the opportunity to begin closing the gap between the reconstruction of the recent past and of the Holocene. Moving between reconstructions of a few hundred years to 1000 years has long been identified as a challenge to geomorphology (Brunsden and Thornes, 1979). Passmore and Macklin (2000) provide a neat example of overcoming some of these difficulties by using a combination of radiocarbon, palaeomagnetic, pollen, lichenometric and trace metal measurements with cartographic evidence to examine phases of incision and aggradation in tributaries of the Tyne.

Third, there has been a recent resurgence of interest in notions of landscape sensitivity and geological conservation. In some cases individual facets of landscape, such as Holocene stratigraphic sections or particular landforms, constitute features of interest for conservation purposes. Existing procedures for Earth science conservation involving SSSIs or RIGS (Regionally Important Geological/Geomorphological Sites) are relevant. In other cases, larger-scale landform assemblages will hold interest, either explicitly as a landscape feature that demands management, or implicitly by recognition of the role of geomorphological process systems in influencing habitat characteristics. As the current state of the landscape is dominantly represented by adjustment to interglacial conditions, management intervention may be necessary to retain transient-state landforms or to protect relict features from contemporary disturbance. Practical guidelines for quantifying limits of acceptable change on a protected landscape feature are contingent on an awareness of process–form relationships and the range of probable magnitudes and frequencies of formative events. Appreciation of geomorphological change over the timescale of 1000 years provides a basis for environmental management.

Fourth, the millennial timescale provides an important framework for appreciating the risks associated with geomorphological change and natural hazards, especially extreme events. This can be readily illustrated with reference to the design of flood defences. To be 95% certain that the defences will not fail within a design lifetime of 50 years, it would be necessary to design for the 1000 year event. In the past, dam failures have led to some of the most catastrophic flood events in Britain. As a consequence, the safe design and maintenance of these structures is strictly controlled under the Reservoirs Act 1975. Large raised reservoirs with a capacity of over 25 000 m^3 are placed in four hazard categories, mainly according to the level of downstream development (table 1.1). Category A reservoirs are those where a breach will endanger lives in a community. The minimum design standard for these dams is the 10 000 year flood. Category B reservoirs are those where a breach might endanger lives in a community or result in extensive damage. The minimum design standard for these dams is the 1000 year flood. The relationship between the probability of a potentially damaging event (often expressed as a return period) and the lifetime of a development or structure also illustrates the need to be aware of millennial-scale changes in the landscape, especially for high-value or high-risk (i.e. those where the consequence of failure would be unacceptable) projects. The probability of a 1000 year event (annual

Table 1.1 Reservoir flood and wave standards by dam category (after Institution of Civil Engineers, 1978)

Category	Initial reservoir condition	Dam design flood inflow		
		General standard	Minimum standard if rare overtopping is tolerable	Alternative standard if economic study is warranted
A. Reservoirs where a breach will endanger lives in a community	Spilling long-term average daily inflow	Probable Maximum Flood (PMF)	0.5 PMF or 10 000 year flood (take larger)	Not applicable
B. Reservoirs where a breach: (i) may endanger lives not in a community; (ii) will result in extensive damage	Just full (i.e. no spill)	0.5 PMF or 10 000 year flood (take larger)	0.3 PMF or 1000 year flood (take larger)	Flood with probability that minimizes spillway plus damage costs; inflow not to be less than minimum standard but may exceed general standard
C. Reservoirs where a breach will post negligible risk to life and cause limited damage	Just full (i.e. no spill)	0.3 PMF or 1000 year flood (take larger)	0.2 PMF or 150 year flood (take larger)	
D. Special cases where no loss of life can be foreseen as a result of a breach and very limited additional flood damage will be caused	Spilling long-term average daily inflow	0.2 PMF or 150 year flood	Not applicable	Not applicable

Table 1.2 The percentage probability of the N-year flood occurring in a particular period

Number of years in period	N = average return period, T_r, in years							
	5	10	20	50	100	200	500	1000
1	20	10	5	2	1	0.5	0.2	0.1
5	67	41	23	10	4	2	1	0.5
10	89	65	40	18	10	5	2	1
30	99	95	79	45	26	14	6	3
60	–	98	95	70	31	26	11	6
100	–	99.9	99.4	87	65	39	18	**9**
300	–	–	–	99.8	95	78	45	26
600	–	–	–	–	99.8	95	70	45
1000	–	–	–	–	–	99.3	87	64

Where no figure is inserted, the percentage probability < 99.9.

In bold type: there is, on average, a 9% chance that a 1000 year event (annual probability of 0.001) would occur within a 100 year time period.

probability of 0.001) occurring within 'engineering time' (generally recognized to be 50–150 years) is around 10% (table 1.2). Thus, the potential for major landslide events, coastal change or river channel migration and so on needs to be taken into account in the planning and design of major engineering projects. The content of the volume serves, therefore, to draw attention to the nature and scale of geomorphology-related issues that might have an influence on a project in a particular location. Forewarned is forearmed.

Fifth, the millennial scale provides an opportunity for developing scenarios for the possible impacts of climate change and sea-level rise. Indeed, it is possible to match climatic records over the last 300–400 years with documentary accounts for major events and early map sources. Within this period there have been notable variations in storminess, not only from year to year but also over decades and centuries. Analysis of the number of storms of different severity class since 1570 reveals marked periods of increased storminess: prior to 1650, between 1880 and 1900, and since 1950 (Lamb, 1991).

The storminess prior to 1650 is widely believed to reflect the period of colder climate known as the 'Little Ice Age'. This period was characterized by frequent severe winters, reduced run-off – Thom and Ledger (1976) suggest that run-off was 89% of present levels – and the occurrence of surface winds of strengths unparalleled in this century. Indeed, most of the major wind-blown sand events are from this period (e.g. the Culbin Sands disaster of 1694 and the Breckland storms of between 1570 and 1588). The end of the Little Ice Age was marked by a wetter, more extreme and variable climate, which may offer an analogue to the current phase of atmospheric warming (Newson and Lewin, 1991). This period from 1700 to 1850 has been associated with marked increases in autumn and winter floods, as identified in north-eastern England (Archer, 1987); major river channel changes, in both the uplands of northern England and in floodplain locations (Macklin et al., 1992); an increase in the reported incidence of major coastal landslides in southern England, such as the 1810 landslip on the Isle of Wight and the Great Bindon landslip of 1839 (Jones and Lee, 1994); and an increase in debris flow activity in the Highlands of Scotland (Innes, 1983). At the time of writing, much of southern Britain is gripped by severe floods that have re-awoken political and media concern with climate change, flood defence policy and planning. The 1000 year perspective on geomorphological process and event magnitude has much relevance for development objectives.

1.3 Evidence for Geomorphological and Climate Change

1.3.1 Inheritance and legacy

The span of 1000 years is insufficient for substantial geomorphological change or landform development except in particularly active or sensitive environments. In a regime of predominantly low-intensity processes following a transition from glacial to interglacial conditions, the rate of modification of landforms is generally low. The effectiveness of geomorphological processes can be considered as a function of the geological structure (the rock type and faulting patterns); the geological history (tectonics and changes in relative relief); past geomorphological processes (Pleistocene oscillations, changes in sea level and the aftermath of glaciation); present geomorphological processes (climate and role of extreme events); and human impact (in terms of deliberate intervention and the indirect consequences of land use change). The geomorphology of Britain is developed on a complex but essentially ancient geological structure, which has seen negligible tectonic activity in recent times. Past geomorphological processes have been considerably more active than the present-day regime. At the maximum extent, Pleistocene ice sheets reached as far south as the north Devon coast and the approximate line between the Severn and Thames estuaries. In areas that were actively glaciated, the impacts are profound but spatially variable. Even in an area such as the Lake District, where there are many classic landforms of mountain glaciation, many remnants of older surfaces can be seen. South of the maximum glacial limits, periglacial processes accomplished much landform change through enhanced weathering and solifluction. The legacy of the transition from glacial to interglacial conditions has been the assemblage of characteristic landforms inherited from past geomorphological process environments and an abundant supply of unstable and unconsolidated sediments. The term 'paraglacial' has been applied to environments that are directly conditioned by glaciation (Church and Ryder, 1972). The 'nearness' of glacial influence can be considered both in space (distance from active glaciation) and time (elapsed since deglaciation). Paraglacial conditions lead to the formation of distinctive landform assemblages in previously glaciated upland environments, such as alluvial fans and debris cones, where the abundant sediment supply, limited vegetation cover and energy input from precipitation permit rapid rates of sediment transfer. On the less dramatic relief of southern England, paraglacial conditions facilitated many mass movements. The transition towards the full interglacial is therefore characterized by the depletion of sediment supplies and the establishment of a vegetation cover that offers some resistance to sediment transfer.

Geomorphological process activity through the Holocene in Britain can be considered in terms of the interplay between the energy input for sediment transfer and the availability of sediment to be transported. The former is fundamentally a function of climate, while the latter is related to both the depletion of the legacy of glacial sediment supply and the role of vegetation cover in inhibiting transfer. Deterioration of the climate to wetter and/or stormier conditions and the role of humans in disturbing the natural vegetation cover or generating fresh sediment supplies are the main themes for a consideration of geomorphological process activity within the Holocene. The relative influence of climate change and of human activity has been a favoured, but distracting, theme (Ballantyne, 1991). The Holocene has traditionally been subdivided by the Blytt–Serander climate scheme into five periods that reflect changing European pollen assemblages. These are the Pre-Boreal (IV, warm, dry and birch-dominated); the Boreal (V/VI, warm, dry and pine-dominated); the Atlantic (VIIa, warm and wet, with the spread of oak and elm); the Sub-Boreal (VIIb, warm and dry); and the Sub-Atlantic (VII, cool, wet and characterized by the onset of forest clearance and the elm decline). Not surprisingly, as further research has investigated Holocene climate and vegetation change, the applicability and synchronicity of the scheme has been called into question. Nevertheless, it is helpful to superimpose the evidence for human impact on the British landscape on to the climate change record in setting the scene for the last millennium. The action of Neolithic farmers (in the early Sub-Boreal) had already reduced woodland locally, especially on the sandy and calcareous soils of the south, but by the start of the Bronze Age, Britain was still largely forested. Radiocarbon dating of pollen assemblages for the margins of the Pennine uplands indicates initial clearances between about 3900 and 3400 BP (Tinsley, with Grigson, 1978). Podsolization, declining fertility and soil erosion were consequences of some clearances. Between 2750 and 2500 BP, climate conditions deteriorated in the transition towards the Sub-Atlantic Period, with a decrease in temperature and increasing wetness. As upland settlement retreated, woodland regeneration was limited in many places, and the heath and bog communities developed on land that had been used previously for grazing. By the onset of the last millennium, much regional forest clearance had been accomplished, and the combined affect of climate deterioration and earlier deforestation had led to irreversible changes in soil properties and natural vegetation communities. The continuing adjustment to the glacial–interglacial transition, the legacy of glacial sediment supply and its reworking and depletion, Holocene fluctuation of climate and the increasing influence of humans set the scene for the operation and effectiveness of geomorphological processes in the most recent 1000 years. The evidence for climate change during this period is summarized below.

1.3.2   The climate of the last millennium

In the final years of the first millennium AD, a number of Norse colonies in Greenland and Newfoundland were established. The expansion of agriculture into the British uplands by Norse farmers a few decades later supports, but does not prove, the hypothesis that climate was markedly warmer at this time. The long-established view is that first few centuries of the last millennium were characterized by warm conditions that have become known as the 'Little Optimum' (Lamb, 1977) or the 'Medieval Warm Period'. This was followed by a deterioration to colder and stormier conditions from the fourteenth century. The cold conditions experienced between the Middle Ages and the mid-nineteenth century have become known as the 'Little Ice Age' (Grove, 1988). As more detailed information about environmental reconstruction is assembled, it has become apparent that there is some considerable spatial and temporal heterogeneity in the conditions experienced. Much of the debate has been built upon the pioneering efforts of Hubert Lamb and of Gordon Manley to characterize past climate.

Lamb's index of medieval climate is based on a wide range of evidence for a warm epoch, which was used to establish relative changes in seasonal climate (such as winter severity, and summer wetness/dryness) on a decadal scale (Lamb, 1965). The Central England Temperature (CET) record, devised by Gordon Manley, is a homogenous temperature series based on a three-station average from Lancashire and the south Midlands. When published (Manley, 1974), it provided a monthly record back to 1659, but it has subsequently been upgraded to a daily basis from 1772 (Parker et al., 1992), updated and corrected for some data discrepancies. The CET has become the most studied climate record in the world (Jones and Hulme, 1997), from which fluctuations and trends can be derived. Present annual temperatures are about 0.7°C warmer than at the start of the record, which is close to the nadir of the Little Ice Age.

Evidence about the British climate during the last 1000 years, from which the indices described above are derived, comes from three types of source – the instrumental record, documentary sources and proxy data (Lamb, 1982). The invention of the thermometer at the turn of the seventeenth century made it possible for the direct recording of weather to begin, and some intact temperature records are available from various European stations from the early eighteenth century. The longest temperature records, as mentioned above, are from Central England. However, systematic procedures for recording daily weather were not widespread until the mid-nineteenth century. Measurements of rainfall exist from the eighteenth century, but again the procedures for measurement were not standardized until more recently. The England and Wales rainfall series has been reconstructed back to 1766 (Wigley and Jones, 1987), but the number of gauges available before the 1830s is limited. The direct record, therefore, provides information on only a fraction of the last 1000 years.

Inferences about weather prior to the instrumental record can be drawn from documentary sources. Here, compilations of weather events by Short (1749), Lowe (1870) and Mossman (1898) are of obvious value. The accounts of eyewitnesses such as Defoe (1704) and Samuel Pepys provide an indication of the scale and impact of great storms and floods. The earliest known diary specifically relating to weather observations was compiled by William Merle, in Oxford, for the seven-year period from January 1337 (Ogilvie and Farmer, 1997). Chronicles, charters and ecclesiastical histories provide some qualitative evidence, although some may have been compiled long after the event. Manorial account rolls are particularly useful from 1300 and make reference to events that affected agricultural productivity. The interpretation of weather records from medieval documents is fraught with a number of difficulties, principally concerning the precision of the dates and the reliability and representativeness of the observations. The switch from the Julian to the Gregorian Calendar, adopted in England in September 1752, necessitates the addition of a few days to pre-1752 dates to make inter-annual comparisons. More often, however, the timing of a meteorological event is described more loosely by the season. The term 'autumn' normally refers to the harvest season and 'spring' to the regeneration of biological growth, such that neither can be regarded as a fixed period when comparing records (Ogilvie and Farmer, 1997). The representativeness of records about the weather that are derived from medieval documents also requires consideration. In manorial account rolls, the weather is generally only mentioned when it has some impact on agricultural yield. There is a bias towards more records for the seasons of summer and autumn when an impact on agriculture might be observed and through a lack of commentary on 'normal' weather. Ogilvie and Farmer (1997) note that a large proportion of English summers between AD 1220 and 1430 were described as dry, which may suggest some over-exaggeration in the reporting and/or the onset of a longer-term period of lower rainfall. In general, records of extreme seasons can only be corroborated when they are mentioned by a number of different manors. The screening of medieval documents to generate reliable information about past weather conditions is a time-consuming process. It is unfortunate, although not unexpected, that many of the pioneering attempts to develop indices of medieval climate, including the Lamb index, built upon existing sources of weather information. The uncritical use of secondary sources has led to a number of misinterpretations being included, and it is for this reason that the clear signals of a Medieval Warm Period followed by the Little Ice Age are rather more complex than at first imagined.

A third source of evidence about the climate comes from proxy data. There are various physical and biological indicators that are suitable for Holocene palaeoclimatological reconstruction. The physical evidence of glacier retreat and advance is not applicable to Britain in the last 1000 years, but reconstructions in Norway and the European Alps indicate marked oscillations throughout the last millennium (Grove, 1988). The oxygen isotope record from high-latitude or high-altitude ice cores can be interpreted as an indicator of temperature at the site. A good correlation has been noted between the Crête, Greenland $\delta^{18}O$ record and the Lamb Index of English temperatures (Dansgaard et al., 1975). Ice cores from Peru and the South Pole also indicate relatively warmer phases in the early centuries of the last millennium and relatively cool phases between the seventeenth and nineteenth centuries (Hughes and Diaz, 1994), but this pattern is not coincident in all ice cores nor does it provide unequivocal evidence of climate in Britain. Of biological indicators, the tree ring record (dendrochronology) offers the prospect of precise dating of events. Chronologies for England and Ireland are based on oaks (Baillie, 1995). In addition to dating wood remnants or timber preserved in archaeological contexts, tree ring patterns are indicative of past phenomena that have impaired growth (such as fire scars or volcanic dust veils) and can be used to infer climate variables from growth rates, relative abundance, episodes of growth initiation or tree death. The archaeological record also provides evidence of episodic phases of building that indicate variable rates of deforestation. The mid-fourteenth century, for example, marks a hiatus of building activity in Britain, assumed to result from the Black Death epidemic (Baillie, 1995). Pollen, plant macrofossil and diatoms can be used to infer climatic conditions, although the relationship between the occurrence of the indicator and the character of the environment does not preclude non-climatic influences. The pollen assemblages that indicate the character of surrounding vegetation may be used to infer changes in temperature and precipitation, but are more likely to reflect anthropogenic impacts. Diatoms, preserved in lake sediments, respond to water chemistry, from which changing catchment hydrology and water quality can be inferred. Plant macrofossil remains in peat bogs have been used to reconstruct changing surface wetness, which in turn can be related to periods of desiccation or to enhanced drainage as the bog surface is incised and eroded (Tallis, 1997). Proxy indicators offer enormous scope for environmental reconstruction but the difficulty of separating climate-induced transitions from land use change is apparent.

As more information about the environment of the last millennium has been compiled, it has become clear that the traditional distinction between a Medieval Warm Period and the Little Ice Age is open for reinterpretation. A generalized sequence of climatic phases is presented in table 1.3. There is a general agreement among climate historians that the opening centuries of the last millennium were warm, perhaps as warm as anything experienced in postglacial times. The Fennoscandian tree ring record (Briffa et al., 1992), glacier fluctuations of the Swiss Alps (Grove and Switsur, 1994) and the British documentary record (Ogilvie and Farmer, 1997) suggest that much of northern and western Europe enjoyed a drying and warming trend from AD 1200, although conditions in Iceland and Greenland appear to have become more severe at this time. Cooling is apparent from the mid-thirteenth to the mid-fourteenth century, with a brief return to warmer conditions in the first half of the sixteenth century. Thereafter, conditions were cooler until the mid-nineteenth century. Even within the warmer phases

Table 1.3 The generalized climatic characteristics of the last millennium. The division, although somewhat arbitrary, indicates that conditions did vary significantly between and within the traditional Medieval Warm Period – Little Ice Age classification. The timing of these phases was not necessarily synchronous or as marked across all parts of Britain

Date (approximate, AD)	Principal climatic characteristics
900–1250	Medieval Warm Period. Higher temperatures with drier summers and slightly wetter winters. Higher rates of evapotranspiration, reduced surface wetness and low snowfall, fewer snowmelt events
1250–1420	Climatic deterioration from the relatively warm and dry MWP with some notably wet summers. The glacier advance in the Alps by 1350 is often considered as the start of the 'Little Ice Age' (Grove and Switsur, 1994)
1420–1470	A cool and damp period with lower temperatures, higher rainfall totals and increased storminess (Lamb, 1982). Several cold winters and failed harvests. Some abandonment of marginal settlements
1470–1550	Climatic amelioration – a brief return to warmer conditions
1550–1700	The nadir of the Little Ice Age (the most recent Holocene neoglaciation). Cold, damp and stormy, particularly during 1550–1610 and 1670–1700, with a less cold period between. Annual temperatures 1.5°C below MWP. Cold autumns and severe winters with increased snowfall. A high incidence of landslides
1700–1740	Climatic amelioration. The 1730s were markedly warm
1740–1850	A cool period. Cool autumns and cold winters with enhanced snowfall. Many river basins experienced episodes of severe flooding in late eighteenth century (Rumsby and Macklin, 1996) and a high incidence of reported landslides (Jones, chapter 3)
1850–2000	The start of the current warming phase. An overall rise in temperature, but with cooler/wetter phases during the late nineteenth century and from the 1950s to the 1970s. Increasing summer dryness and warmer and wetter winters in the last two decades of the twentieth century

of the Medieval Warm Period, there are documentary records of severe winters – 1205, for example, has an early report of the Thames being frozen. Other years, such as 1314 and 1318, had markedly wet summers that led to widespread crop failures (Ogilvie and Farmer, 1997). Neither is the evidence for the timing and severity of the Little Ice Age clear-cut and distinct. Alpine glaciers had already occupied advanced positions by AD 1350 (Grove and Switsur, 1994) before receding until the seventeenth century. Although the CET record indicates that temperatures have risen since the mid-seventeenth century, there are several fluctuations. The coldest periods were from about 1560 to 1610 and from 1670 to 1700, when mean annual temperatures were depressed by 1.4–1.7°C compared with the preceding Medieval Warm Period (Lamb, 1985). The autumns were cold and the winters particularly severe. Commencing with the 'Great Winter' of 1564–5, there were frequent 'Great Frosts', accompanied by the widespread freezing of lowland rivers, conditions that have become immortalized by the landscape paintings of the time and in the culture of Christmas. Average annual snow lie was 20–30 days in lowland areas in 1670–1700. The winter of 1683–4 is the coldest on record and 1740 has the lowest annual temperature (Jones and Hulme, 1997). The last two decades of the seventeenth century appear to have been exceptionally wet, as there are several records of floods and crop failures from this period. But within the so-called Little Ice Age, the 1730s and 1830s were markedly warmer decades (Parry, 1978). Nevertheless, recent Northern Hemisphere reconstructions confirm the persistence of cooler conditions back to AD 1500 equivalent to a temperature reduction between 0.5 and 1°C (Mann et al., 1998; Huang et al., 2000).

These particularly cold episodes were flanked by two periods of lesser cooling, the first lasting from 1420 to 1470 – when cool winters and a succession of 'failed summers' led to some settlements being abandoned or 'deserted' – and from 1740 to 1900, which is characterized by cool autumns and by depressed winter temperatures with relatively large numbers of 'cold' days prior to 1850. The evidence from northern parts of Britain points to a broadly similar pattern of change, except that the severity of the reduction in temperatures during cold episodes appears to have been greater. Thus temperatures in the Scottish Lowlands during the period 1670–99 are thought to have been 2–3°C lower than the average recorded for 1930–59, as compared with an equivalent lowering of 1.0–1.1°C in central England (Lamb, 1985). As a consequence, there are many reports of permanent snowfields on the crests of the Cairngorms (1200–1300 m) between 1805 and 1823, thereby pointing to their likely increased extent and persistence in earlier times.

The direct influence of temperature on geomorphological processes is limited, albeit that it may influence the degree of freeze–thaw activity that may be important for mechanical weathering, upland slope processes and bank erosion. Indirectly, the influence of temperature on agricultural activity accounts for land use change that has potential geomorphological consequences. Parry (1978), for example, has demonstrated the substantial amount of abandoned arable land in the Southern Uplands of Scotland. Over 20% of the existing moorland of the Lammermuir Hills has evidence of former cultivation. It is likely that colonization occurred in the eleventh and twelfth centuries following the introduction of the moldboard plough, but reduced temperature and increased wetness from the fourteenth century onwards made some of this land sub-marginal. Several settlements were abandoned in the seventeenth century. As noted above, the 1690s were particularly severe, and are referred to in Scottish folklore as the Seven Ill Years (Whyte, 1981). Conditions for the people were bleak and the cultivation limits retreated downslope by as much 200 m (Parry, 1978). From a process viewpoint, information about precipitation is more important than the temperature record. Attempts to reconstruct British rainfall patterns since the mid-eighteenth century (Wigley and Jones, 1987) show little evidence of systematic trends. Seasonal precipitation totals are highly variable from year to year, but there is some evidence that winters have become wetter and summers drier (Jones and Conway, 1997; Jones et al., 1997), particularly in the last 25 years. The early instrumental records show that much of Europe experienced dry decades between 1730 and 1760 but relatively wet years between 1760 and 1780 (Jones and Bradley, 1992). Estimates of pre-instrumental precipitation totals suggest that the Medieval Warm Period experienced drier summers and wetter winters, with annual precipitation up by about 3%, while the Little Ice Age had decreases of around 7–10% against the 1916–50 mean (Lawler, 1987). The implication of episodic climate change for geomorphological response is considered further in chapter 2, but it is clear that decadal and annual variations mask the potential influence of short-term events (individual storms or floods) that are capable of accomplishing much geomorphological change. Higher-resolution climate reconstruction is possible using schemes such as the Lamb Classification of Daily Weather Types. The classification is based on the direction of general air flow and the motion of synoptic systems, and covers the period from 1861 to the present. Wilby et al. (1997) report strong correlation between the frequency of winter cyclonic Lamb weather type and variations in lake-based sediment yields.

1.3.3   The historical record of geomorphological change

Much of our knowledge of the geomorphological changes over the last 1000 years relies on the historical archive of public records, journals, diaries, newspapers, photographs, maps and charts. However, the archive does not provide a consistent or unbiased record of the last 1000 years, either from a temporal or process/event perspective. Where pronounced or noticeable changes have occurred in the landscape, they may have been recorded on maps or charts (Hooke and Kain 1982; Hooke and Redmond, 1989). So-called 'county' maps date from Elizabethan times to the nineteenth century and are of variable scale and quality. Carr (1969), for example, describes their use in defining coastal changes at Orford Ness. In Scotland, Roy's Military Survey of Scotland (1745–55) provides a fairly accurate record at 1 inch to 1000 yards (1 : 36 000 scale). The earliest manuscript maps (estate maps, tithe or enclosure maps) date from the late eighteenth and early nineteenth centuries. They vary considerably in scale and standard of cartography. Detailed scale (3–6 chains to the inch – between 1 : 2376 and 1 : 4752 scale) tithe maps were produced for 75% of England and Wales between 1838 and 1845, in compliance with the Tithe Commutation Act of 1836. The first Ordnance Survey maps were produced at 1 inch to the mile (1 : 63 360 scale) between 1805 and 1873, but are considered to be of dubious accuracy (Carr, 1962; Harley, 1968). Larger-scale, 1 : 2 500 and 1 : 10 560 maps were produced from the 1870s onwards. The number of subsequent editions of these maps depends largely on the amount of development in an area.

Analysis of change from historical maps is not without its problems. Although the positional accuracy of many defined objects on Ordnance Survey maps is estimated to be ± 0.8 m, inaccessible features of 'marginal importance' situated away from settlements may not be mapped with comparable accuracy (Carr, 1962, 1980). Often, landscape features of interest are not clearly defined in the field, and their map position may be based on a surveyor's perception of their form. As a result, plotting on different editions or different sheets of the same edition may be sensitive to operator variance (Hooke and Kain, 1982). Not all features are revised for each new map edition, so it is sometimes uncertain exactly when a particular feature was last revised.

Significant geomorphological events may have been recorded in journals or diaries. However, these sources provide only a very limited picture of the first half of the millennium. It is possible to use such sources to define the major flood events on the larger rivers. On the Thames, for example, the Anglo-Saxon Chronicles record severe flooding on the East Coast in 1099. However, probably the earliest reliable record of a landslide event dates from 1571, near Kynaston, Herefordshire:

> on the 17th of February, at six o-clock in the Evening, the Earth began to open, and a Hill with a Rock under it . . . lifted itself up a great height, and began to travel, bearing along with it the Trees that grew upon it, the Sheep folds and Flocks of Sheep abiding there at the same time. In the place from where it was first mov'd it left a gaping distance forty foot broad, and fourscore ells long; the whole Field was about 20 acres. Passing along, it overthrew a Chapel standing in the way . . . (Baker, 1674).

Over the last 150–200 years, newspapers have become an important vehicle for recording dramatic events. For example, the major floods of the nineteenth century, such as those of October 1875, were described in considerable detail, sometimes occupying almost entire newspapers. Often, the accounts may contain references to previous similar events. As an example, the following summarizes an article appearing in the *Nottingham Daily Guardian* for 2 January 1901:

> A heavy and prolonged downpour on Sunday, 30 December 1900. It was recorded locally as 1.684 inches, and compared with the year's previous heavy fall of 1.286 inches on June 11 1900. The water-level rose rapidly during the night of Monday 1 Jan/Tuesday 2 Jan 1901, and by the morning of the 2nd flooding was taking place over both banks of the river. By Tuesday night the flood level had passed the 1869 flood mark and had almost reached the 1852 flood mark.

The value of local newspaper sources has been highlighted by Lee and Moore (1991), who established the pattern of contemporary ground movement in the Ventnor landslide complex, Isle of Wight, from a systematic search of local newspapers from 1855 to the present day. The search identified over 200 individual incidents of ground movement and allowed a detailed model of landslide potential to be developed that formed the basis for planning and management. More subtle changes or less dramatic events in isolated areas are not generally recorded in the historical archive. Thus, while we might know something of the suspected increase in major landslide activity during the Little Ice Age (Jones and Lee, 1994), we know relatively little about hillslope erosion, floodplain accretion, saltmarsh growth and other less spectacular processes throughout this period. Here, our knowledge is limited to sedimentological evidence (with associated problems of dating) and very recent (i.e. within the last 30 years) scientific monitoring data or chance records.

Finally, in attempting to examine geomorphological change in Britain during the last 1000 years, geographical variability should not be overlooked. The temptation to link reconstructed geomorphological events between regions, or even within catchments, requires caution. The incidence of upland gully erosion and renewed aggradation of alluvial fans in the Howgill Fells (Harvey et al., 1981) and the Forest of Bowland (Harvey and Renwick, 1987) have been radiocarbon dated to the early eleventh century, a time when Viking settlers were colonizing the upland margins. A similar timing of fan aggradation is apparent at sites in the Southern Uplands of Scotland (Tipping and Halliday, 1994). These upland examples of instability have recently been accompanied by evidence of a major soil erosion event and deposition of clastic material in Slapton Ley on the South Devon coast (Foster et al., 2000). The silty-clay unit is dated at two locations at 910 ± 160 and 960 ± 140 BP (conventional radiocarbon ages) and is evident in the floodplain sediments of other Devon rivers. Any millenarian looking for signs that the transition between the first and second millennium AD was afflicted with wrathful landscape instability might find solace in this apparent coincidence in timing across distant parts of Britain. Furthermore, the non-materialization of the end of the world at AD 1000 was excused by some millenarian clerics of the age to indicate that the Apocalypse would be unleashed when 1000 years had elapsed since Christ's passion rather than His nativity (Gould, 1998). Not only did the requisite year of AD 1033 have reports of widespread famine in Europe, but it fits neatly within the radiocarbon age ranges for the various archives of instability identified in Britain. However, it should be remembered that age ranges are indeed ranges. Tipping and Holliday (1994) have demonstrated through careful stratigraphical analysis that deposits in the upper Tweed valley, although broadly synchronous in age, relate to more than one event. There is an ongoing challenge to improve the precision of dating and further develop techniques of palaeoenvironmental reconstruction. This will enable investigations to move beyond general inferences about the temporal coincidence of possible causal effects towards more subtle questions about the mechanisms and timing of geomorphological process activity.

1.4   The Structure of the Book

The ten chapters of this book evaluate the evidence for geomorphological process activity over the last 1000 years, the techniques that are suitable to decipher such activity and the implications for both landscape development and the management of the landscape. As such, the authors draw on material from the published literature, consultancy reports and their own research projects. Throughout the book, calendar dates are referred to as AD and general periods of time in terms of centuries. Where evidence is deduced from radiocarbon dates, these are usually reported as BP (before present, AD 1950). Radiocarbon dates should be assumed to be conventional (i.e. uncalibrated) dates, unless specifically indicated. The authors were given specific subject briefs, but our definition of Britain has remained somewhat fluid. It is not the intention to provide an exhaustive geographical coverage of landscape change but, rather, to illustrate the principles through well chosen examples and case studies. Some geographical bias within individual chapters may reflect the research histories of the authors, but some care has been taken to ensure that examples are provided from England, Scotland Wales and Ireland. The authors also use examples from overseas temperate environments as analogues.

Denys Brunsden (chapter 2) provides some context by considering the boundary conditions for geomorphological activity over a 1000 year timescale. Invited to address the International Association of Geomorphologists, Brunsden (1990) defined a set of ten propositions about geomorphological behaviour that have earned wide circulation and some notoriety. These propositions are employed as a guiding framework to ask what change is possible over a timescale of a millennium. The curious main title is a Dorset dialect expression that captivates the essence of a living history. Chapters 3–6 are framed by the sediment cascade that considers geomorphological process activity from source to sink. David Jones (chapter 3) considers hillslope processes during the last 1000 years, with particular emphasis on the ways in which human activity has increased the propensity for erosion. Estimates of anthropogenic redistribution of earth surface materials suggest that geomorphological processes have become much less significant than humans in remoulding the British landscape. Having delivered some sediment from slopes to valley floors, Barbara Rumsby (chapter 4) evaluates the significance of valley-floor and floodplain processes, and identifies two major phases of enhanced fluvial activity in the eleventh and late eighteenth centuries. Developments in techniques to reconstruct process environments are covered in this chapter. Janet Hooke (chapter 5) focuses on the evidence for river channel change and provides a number of case studies of millennial-scale reconstruction. The extent and early history of modification in British river channels becomes apparent and yet the spatial variability in response is marked. Moving downstream, Mark Lee (chapter 6) investigates estuarine and coastal processes. Whereas many inland areas have remained insensitive to change during the last 1000 years, the British coast is a dynamic environment that exhibits much change and plenty of opportunities for applied geomorphology, as partly reflected in the large number of consultancy reports from which data are derived. In order to emphasize the importance of linkages in the sediment delivery system, two further chapters consider sediment transfer. David Higgitt, Jeff Warburton and Martin Evans (chapter 7) examine upland sediment transfer, reviewing attempts to develop chronologies for significant events in upland environments and to quantify upland sediment yields. Ian Foster (chapter 8) provides the complementary review for lowland rural environments, focusing on the identification of retention mechanisms, conveyance losses and transfer processes. Although understanding of sediment transfer mechanisms and pathways has improved in recent years, in both upland and lowland environments, many questions remain unanswered.

With the exception of coastal systems and some sensitive fluvial environments, process activity in Britain over the last 1000 years has been relatively modest. That is not to say that geomorphological expertise does not have relevance in many engineering situations. Mark Lee (chapter 9) reviews the various ways in which geomorphological processes present management issues for the effective use of land. There is a long history of economic and human loss associated with natural hazards, particularly flooding, which raises many questions concerning future land development and the prospects of climate change. The volume ends with a short concluding chapter by the editors, that focuses on the impact of humans on the British environment over the last 1000 years and the challenges that such an interaction raises for geomorphology.

REFERENCES

Archer, D. R. 1987: Improvement in flood estimates using historical information on the River Wear at Durham. National Hydrology Symposium, Hull.
Archer, D. 1992: *Land of Singing Waters: Rivers and Great Floods of Northumbria*. Stocksfield: Spedden Press.

Baillie, M. G. L. 1995 *A Slice Through Time: Dendrochronology and Precision Dating*. London: Batsford.

Ballantyne, C. K. 1991: Late Holocene erosion in upland Britain: climatic deterioration or human influence. *The Holocene*, 5, 25–33.

Baker, Sir Richard 1674: *A Chronicle of the Kings of England etc.*, 6th impression. London, Ludgate-hill: George Sawbridge, Hosier Lane.

Briffa, K. R., Jones, P. D., Bartholin, T. S., Eckstein, D., Schweingruber, F. H., Karlen, W., Zetterberg, P. and Eronen, M. 1992: Fennoscandian summers from AD 500: temperature changes on short and long timescales. *Climate Dynamics*, 7, 111–19.

Brunsden, D. 1990: Tablets of stone: toward the ten commandments of geomorphology. *Zeitschrift für Geomorphologie, Supplementband* 79, 1–37.

Brunsden, D. and Thornes, J. B. 1979: Landscape sensitivity and change. *Transactions of the Institute of British Geographers*, NS, 4, 303–484.

Carr, A. P. 1962: Cartographic error and historical accuracy. *Geography*, 47, 135–44.

Carr, A. P. 1969: The growth of Orford Spit: cartographic and historical evidence from the sixteenth century. *Geographical Journal*, 135, 28–39.

Carr, A. P. 1980: The significance of cartographic sources in determining coastal change. In R. A. Cullingford, D. A. Davidson and J. Lewin (eds), *Timescales in Geomorphology*. Chichester: Wiley, 67–78.

Church, M. A. and Ryder, J. M. 1972: Paraglacial sedimentation: a consideration of fluvial processes conditioned by glaciation. *Geological Society of America Bulletin*, 83, 3059–72.

Coones, P. and Patten, J. 1986: *The Penguin Guide to the Landscape of England and Wales*. Harmondsworth: Penguin.

Dansgaard, W., Johnsen, S., Reeh, N., Gundestrup, N., Clausen, H. and Hammer, C. 1975: Climatic changes, Norsemen and modern man. *Nature*, 255, 24–8.

Defoe, D. 1704. *The Storm*. London.

Foster, I. D. L., Mighall, T. M., Wotton, C., Owens, P. N. and Walling, D. E. 2000: Evidence for mediaeval soil erosion in the South Hams region of Devon, UK. *The Holocene*, 10, 261–71.

Goudie, A. S. and Brunsden, D. 1994: *The Environment of the British Isles: an Atlas*. Oxford: Oxford University Press.

Goudie, A. S. and Gardner, R. 1992: *Discovering Landscape in England and Wales*. London: Chapman and Hall.

Gould, S. J. 1998: *Questioning the Millennium*. London: Vintage.

Grove, J. M. 1988 *The Little Ice Age*. London: Methuen.

Grove, J. M. and Switsur, R. 1994: Glacial geological evidence for the Medieval Warm Period. *Climatic Change*, 26, 143–69.

Hansom, J. 1998: The coastal geomorphology of Scotland: understanding sediment budgets for effective coastal management. In *Scotland's Living Coastline*. Edinburgh: Stationery Office, 34–44.

Harley, J. B. 1968: Error and revision in early Ordnance Survey maps. *Cartographic Journal*, 5, 115–24.

Harvey, A. M. and Renwick, W. H. 1987: Holocene alluvial fan and terrace formation in the Bowland Fells, northwest England. *Earth Surface Processes and Landforms*, 12, 249–57.

Harvey, A. M., Oldfield, F., Baron, A. F. and Pearson, G. W. 1981: Dating of post-glacial landforms in the central Howgills. *Earth Surface Processes and Landforms*, 6, 401–12.

Hooke, J. M. and Kain, R. J. P. 1982: *Historical Change in the Physical Environment: a Guide to Sources and Techniques*. Sevenoaks: Butterworths.

Hooke, J. M. and Redmond, C. E. 1989: Use of cartographic sources for analysing river channel change with examples from Britain. In G. E. Petts, H. Moller and A. L. Roux (eds), *Historical Change of Large Alluvial Rivers: Western Europe*. Chichester: Wiley, 79–94.

Hoskins, W. G. 1955: *The Making of the English Landscape*. London: Hodder and Stoughton.

Huang, S., Pollack, H. N. and Shen, R. Y. 2000: Temperature trends over the past five centuries reconstructed from borehole temperatures. *Nature*, 403, 756–8.

Hughes, M. K. and Diaz, H. F. 1994: Was there a 'Medieval Warm Period', and if so, where and when? *Climate Change*, 26, 109–42.

Innes, J. L. 1983: Lichenometric dating of debris flow deposits in the Scottish Highlands. *Earth Surface Processes and Landforms*, 8, 579–88.

Institution of Civil Engineers 1978: *Floods and Reservoir Safety*. London: Thomas Telford.

Jones, D. K. C. and Lee, E. M. 1994: *Landsliding in Great Britain*. London: HMSO.

Jones, P. D. and Bradley, R. S. 1992: Climatic variations in the longest instrumental records. In R. S. Bradley and P. D. Jones (eds), *Climate Since AD 1500*. London: Routledge, 246–68.

Jones, P. D. and Conway, D. 1997: Precipitation in the British Isles: an analysis of area-average data updated to 1995. *International Journal of Climatology*, 17, 427–38.

Jones, P. D. and Hulme, M. 1997: The changing temperature of 'Central England'. In M. Hulme and E. Barrow (eds), *Climates of the British Isles: Present, Past and Future*. London: Routledge, 173–96.

Jones, P. D., Conway, D. and Briffa, K. R. 1997: Precipitation variability and drought. In M. Hulme and E. Barrow (eds), *Climates of the British Isles: Present, Past and Future*. London: Routledge, 197–219.

Lamb, H. H. 1965: The early medieval warm epoch and its sequel. *Palaeogeography, Palaeoclimatology, Palaeoecology*, 1, 13–37.

Lamb, H. H. 1977: *Climate: Present, Past and Future*. Vol. 2. *Climate History and the Future*. London: Methuen.

Lamb, H. H. 1982: *Climate, History and the Modern World*. London: Methuen.

Lamb, H. H. 1985: Climate and Landscape in the British Isles. In S. R. J. Woodell (ed.), *The English Landscape: Past, Present and Future*. Oxford: Oxford University Press, 148–67.

Lamb, H. H. 1991: *Historic storms of the North Sea, British Isles and Northwest Europe*. Cambridge: Cambridge University Press.

Lawler, D. M. 1987: Climatic change over the last millennium in Central Britain. In K. J. Gregory, J. Lewin and J. B. Thornes (eds), *Palaeohydrology in Practice*. Chichester: Wiley, 99–129.

Lee, E. M. and Moore, R. 1991: *Coastal Landslip Potential Assessment; Isle of Wight Undercliff, Ventnor*. Geomorphological Services Ltd.

Lowe, E. J. 1870: *Natural Phenomena and Chronology of the Seasons*. London.

Macklin, M. G., Rumsby, B. T. and Newson, M. D. 1992: Historic floods and vertical accretion in fine grained alluvium in the Lower Tyne Valley, North East England. In P. Billi, R. D. Hey, C. R. Thorne and P. Tacconi (eds), *Dynamics of Gravel Bed Rivers*. Chichester: Wiley, 573–89.

Manley, G. 1974: Central England temperatures: monthly means 1659 to 1973. *Quarterly Journal of the Royal Meteorological Society*, 100, 389–405.

Mann, M. E., Bradley, R. S. and Hughes, M. K. 1998: Global-scale temperature patterns and climate forcing over the past six centuries. *Nature*, 392, 779–87.

Mossman, R. C. 1898: The meteorology of Edinburgh, Part III. *Transactions of the Royal Society of Edinburgh*, 40(iii), 476–8.

Mottershead, D. N. 2000: Weathering of coastal defensive structures in southwest England: a 500 year stone durability trial. *Earth Surface Processes and Landforms*, 25, 1143–59.

Newson, M. and Lewin, J. 1991: Climatic change, river flow extremes and fluvial erosion – scenarios for England and Wales. *Progress in Physical Geography*, 15, 1–17.

Ogilvie, A. E. J. and Farmer, G. 1997: Documenting the medieval climate. In M. Hulme and E. Barrow (eds), *Climates of the British Isles: Present, Past and Future*. London: Routledge, 112–33.

Parker, D. E., Legg, T. P. and Folland, C. K. 1992: A new daily Central England temperature series. *International Journal of Climatology*, 12, 317–42.

Parry, M. L. 1978: *Climate Change, Agriculture and Settlement*. Folkestone: Dawson.

Passmore, D. G. and Macklin, M. G. 2000: Late Holocene channel and floodplain development in a wandering gravel-bed river: the River South Tyne at Lambley, Northern England. *Earth Surface Processes and Landforms*, 25, 1237–56.

Pocock, D. C. D. 1996: Place evocation: the Galilee Chapel in Durham Cathedral. *Transactions of the Institute of British Geographers*, NS, 21, 379–86.

Rumsby, B. T. and Macklin, M. G. 1996: River response to the last neoglacial (the 'Little Ice Age') in northern, western and central Europe. In J. Branson, A. G. Brown and K. J. Gregory (eds), *Global Continental Changes: the Context of Palaeohydrology*. Geological Society, London, Special Publication No. 115, 217–33.

Short, T. 1749: *A General Chronological History of the Air, Weather, Seasons, Meteors*. London: Longman.

Tallis, J. H. 1997: The pollen record of *Empetrum nigrum* in southern Pennine peats: implications for erosion and climate change. *Journal of Ecology*, 85, 455–65.

Thom, A. S. and Ledger, D. C. 1976: Rainfall, runoff and climatic change. *Proceedings of the Institution of Civil Engineers*, 61, 633–52.

Tinsley, H. M. with Grigson, C. 1978 The Bronze Age. In I. G. Simmons and M. J. Tooley (eds), *The Environment in British Prehistory*. London: Duckworth, 210–49.

Tipping, R. and Halliday, S. P. 1994: The age of alluvial fan deposition at a site in the southern uplands of Scotland. *Earth Surface Processes and Landforms*, 19, 333–48.

Trueman, A. E. 1949: *Geology and Scenery in England and Wales*. Harmondsworth: Pelican.

Whyte, I. D. 1981: Human response to short- and long-term climatic fluctuations: the example of early Scotland. In M. L. Parry and C. Delano-Smith (eds), *Consequences of Climatic Change*. Department of Geography, University of Nottingham, Nottingham, 17–29.

Whyte, I. D. and Whyte, K. A. 1991: *The Changing Scottish Landscape 1500–1800*. London: Routledge.

Wilby, R. L., Dalgleish, H. Y. and Foster, I. D. L. 1997: The impact of weather patterns on historic and contemporary catchment sediment yields. *Earth Surface Processes and Landforms*, 22, 353–63.

Wigley, T. M. L. and Jones, P. D. 1987: England and Wales precipitation: a discussion of recent changes in variability and an update to 1985. *Journal of Climatology*, 7, 231–46.

Chapter 2

Back A'long: a Millennial Geomorphology

Denys Brunsden

2.1 Introduction

During the last century there has been an impressive increase in information about changes in uplift, sea level, climate and environment, as well as an increasing knowledge of process rates and mechanisms. The quality of our dating techniques means that we have quite a good understanding of the way in which system controls are changing and the direction and scale of some of the responses during the last 10 000 years. Curiously, however, the last 1000 years is not so well known either in a geomorphological or palaeo-ecological sense as we might wish. As Roberts (1998) notes, most scientists looking for ecological or sedimentation change have ignored the last few centimetres of cores and concentrated on information from well before the modern period. Most measured records of processes are still, with the exception of fluvial discharge, on a sub-decade scale. There is sound historical knowledge of human activity covering the last millennium and the archives for the last two centuries provide quite accurate data on mesoscale changes to some of the controls of landscape change in documentary, map and photograph form. Unfortunately, however, they do not document either the scale or location of geomorphological change except in the gross sense of channel or coastline positions. The purpose of this chapter is to examine how well we understand the dynamic basis of the subject across the 1000 year timescale.

The chapter will utilize the propositions suggested as a possible theoretical framework for the subject by Brunsden (1990), despite the comments of Clayton (1997) who regards these 'unhappy' ideas as 'too weak to sustain the attention'. The propositions are used here as a framework because they are not considered by everyone to be trivial (see McCann and Ford, 1996) and because they provide a way to respond to a challenge by Stoddart (1978), 'Can we extrapolate the short term record of measurable processes to the relatively unknown time span of 100–10 000 years and beyond?' Since the brief for the chapter was to cover the 1000 year timescale, it is hoped that the ideas will help to bring some structure to a difficult task. They will be referred to in the text as propositions 1–10 without further references.

2.2   Tectonic and Stress Fabric Control

Propositions 1 and 2 suggest that '*the style and location of landform change is determined by the type, location and rate of tectonic movement and their associated stress fields interacting with denudation processes over the relevant time and space framework of the landform assemblage*' and that landforms '*directly reflect the ratio between the rates of operation of these processes*'.

The tectonic story of Great Britain is long and complex. The setting for our story lies at the junction of three great orogenic belts of Precambrian (1300–900 Ma), Caledonian (500–400 Ma) and Hercynian (300 Ma) age. Mid-Tertiary Alpine movements have affected all the stress fabrics of these formative events, and the patterns produced form the most important underlying feature of the British geomorphological domain at all scales. By the time our story begins, the underlying structures and rock character had already been exploited by the current processes for a long time, but they still form the nuclei on which our millennium story is based (figure 2.1(a)).

On the 1000-year timescale, however, tectonic movement is not a primary forcing function of landform change in Great Britain. Available data (Shennan, 1989) show (figure 2.1(b)) that the postglacial trends of isostatic uplift in Scotland and subsidence in the south-west and south-east continue today. Over the millennium there is a maximum north–south differential of *c*.4 m.

The traditional plots (e.g. Gray 1992), however, should be regarded with caution. While the subsidence in the south-east has an underlying structural rationale in the London Basin and the North Sea Rift, the published maps of isostatic uplift in Scotland usually ignore all fundamental controls. There seems to be an unwritten law that maps of isostatic movements for Scandinavia and northern Canada as well as Scotland should be drawn in a simplified domed form. Yet these areas are all ancient landscapes with major tectonic fractures, sedimentary basins and axes of uplift of great antiquity and influence, and in places the landscape of Britain has a clear tectonic block form. As Sissons

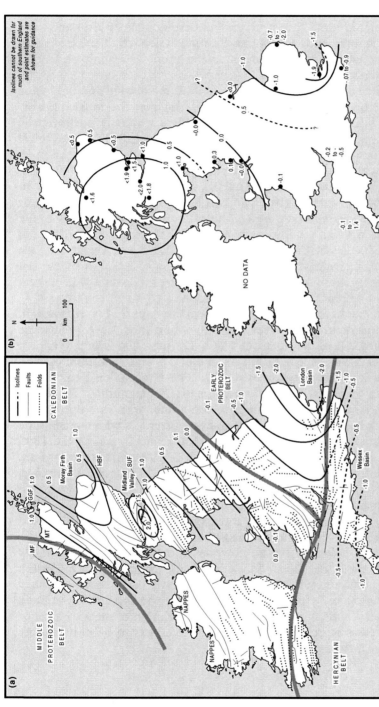

Figure 2.1 (a) A general structural and current uplift map of the British Isles. The Great Glen Fault (GGF), Highland Boundary Fracture Zone (HBF), Lizard Thrust Zone (LT), Minch Fault (MF), Moine Thrust (MT) and Southern Uplands Fault (SUF) are emphasized. The structural provinces shown are the Middle Proterozoic, the Caledonian, the Hercynian and the Early Proterozoic basement. The uplift data from (b) (Shennan, 1989) are reinterpreted to take some account of the main structural blocks. Future models of uplift might refine the interpretation to take account of other structural basins. Uplift rates are in mm yr⁻¹ (which can probably be extrapolated to units of metres in the last 1000 years). (b) Estimated current rates of crustal movement in Great Britain (after Shennan, 1989), in mm year⁻¹. This is the traditional isoline view, which takes no account of the possible independent isostatic response of crustal blocks.

(1992) pointed out, the structural underpinnings of the landscape cannot be ignored and, at the least, the Great Glen, Highland Boundary and Southern Upland faults ought to be used as tectonic and uplift behaviour boundaries.

If the current uplift data (Shennan, 1989) are plotted and only these three structures are included, it will be seen that the major tectonic blocks of Scotland may not have behaved as generally portrayed by textbooks (figure 2.1(a)). For example, the data actually falls into logical groups. The Central Lowlands – not the Highlands – show the main changes, with a maximum uplift rate of 2 mm year or, if we can extrapolate the data to the millennium, 2 m in the last 1000 years. In the same period parts of the Highlands have only changed by 0.5 m or less. There appears to be a general tilt of the rest of the country to the south-east, with an additional component into the ancient London Basin subsidence of the Palaeozoic floor. The South Coast data are unreliable, but suggest a general subsidence of 0.1–1.0 m in the last 1000 years. A thorough re-examination of the uplift data with respect to the building blocks seems to be in order.

Seismic shocks are frequent throughout the country but are small in intensity and magnitude. Over the millennium there have been some 2000 recorded events since AD 1185. During this time no known earthquakes have exceeded a magnitude of 5.5 and maximum epicentral intensities have not exceeded VIII. The Great Glen Fault of the Caledonian orogen still records tremors of 4.5 magnitude. The South Wales–Herefordshire Caledonian tectonic line, with one-third of all UK shocks, has the highest concentration of tremors with a magnitude of >4 located on the Neath–Swansea disturbance, such as the 1896 Hereford shock. Events related to North Sea tectonic structures are well marked, with notable events in 1927 and 1931 east of the Humber, 1382 in North Kent, 1580 near the Straits of Dover (the so-called London quake) and in Colchester in 1884. The latter which, at 5 km depth, was a shallow earthquake at a magnitude of 4.4, was felt 300 km away, killed three people and damaged 1000 houses.

In geomorphological terms, however, these small events probably mean that there has been little identifiable concurrent relationship between uplift and erosion. There may have been some tendency towards coastal and estuarine sedimentation in the north and continued erosion of the cliffs of the south, but the resolution of measurement of these processes is not good enough for the effects to be noticed over the 1000 year timescale.

2.3   Base-level Control

The third proposition for geomorphology is that '*the lower boundary conditions for landform development are set by the varying sea levels experienced during the lifetime of the landscape*'. The context is that the parts of the British landscape that are affected by base-level change during the Holocene are dominated by the 100 m rise during the Holocene (figure 2.2(a)). There has also been significant local subsidence due to extraction of water, salt and minerals especially in Cheshire and near Ripon, Yorkshire (Cooper and Waltham, 1999). These effects far exceed tectonic effects and can have a significant local land-forming influence, including subsidence 'flashes', dolines and enhancement of structural cracks.

The current changes of sea level (figure 2.2(b)) mainly serve to enhance the tectonic forcing. Northern Scotland shows sea-level rise of less than 1 m in the last 1000 years, in the Hebrides and on the East Coast where overall rise is not masked by uplift. The Firth of Forth lies in a zone of falling sea levels (3–7 m), as does the Moray Firth. Levels appear to be falling in Pembrokeshire and Norfolk but the rest of Britain seems to be suffering from King Canute's problem at rates of 1–5 m in the last 1000 years.

2.4   Form Adjustment

Proposition 4, '*For any set of environmental conditions, through the operation of a constant set of processes, there will be a tendency over time to produce a set of characteristic landforms*', is difficult to demonstrate on the 1000 year timescale except at limited scales and for active systems. This is because much of the British landscape is in a relict form from past regimes and there has not been sufficient time to allow adjustment. The processes that are 'characteristic' of the last millennium are those of a quite benign, interglacial, temperate, fluvial process system, in which the rates and intensity of change are not high, modifications of the characteristic forms of the previous regime are slow and relaxation times are very long. Only the most labile systems, such as beaches and eroding cliffs, can be expected completely to reflect current processes and environmental controls.

A further problem is that much of the country has suffered from a very harsh and geomorphologically effective glacial and periglacial regime, during which much of the available work of sediment transport and mass movement was achieved. In lowland Britain many hillslopes

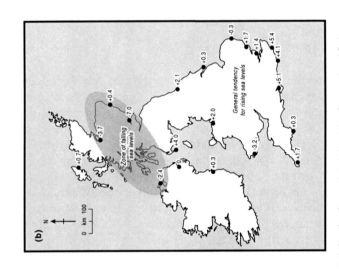

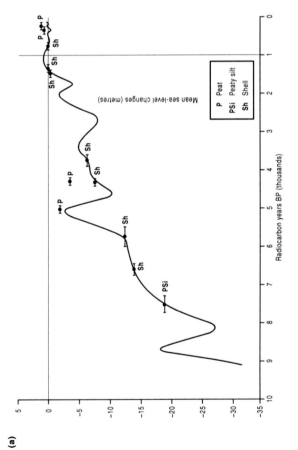

Figure 2.2 (a) Changes in mean sea level during the Holocene at the Essex coast, based on radiocarbon, lithological, faunal and geotechnical data (after Greensmith and Tucker, 1973). The detail on the curve may be exaggerated (Shennan, 1989), but indicates the possible variations in sea level during the last millennium. (b) Recent sea-level changes, in mm yr^{-1} (after Carter, 1988).

were reduced to quite low angles. These are 'unchangeable', in a mechanical sense, by the current processes and are therefore merely 'accepted' as suitable for the conditions. Brunsden and Thomes (1979) have described this as an over-relaxed landscape. In addition, important adjustments were made during the Holocene as the climate changed from glacial to interglacial. This is particularly true of the uplands, where streams and coastal processes worked at overcoming the paraglacial legacy. Completed adjustments, however, are largely restricted to the floodplains, beaches and sensitive areas.

2.5 Episodic Cause

Our understanding of these problems is improved if we take proposition 5 into account, namely that, '*Landforms are continually subject to perturbations which arise from changes in the environmental conditions of the system. These impulses are episodic and complex in nature at all scales. Therefore changes to landforms will be episodic and complex.*' As Schumm (1976) has shown, perturbations can also be caused by progressive change and exceeding of thresholds in the system.

2.5.1 Natural causes

In addition to the tectonic and base-level controls, landforms respond to the rhythms and thresholds of the earth as transmitted over time by the climatic and environmental change processes (figure 2.3). Over a 10^6 year period, the best known of these are the long-term trends associated with astronomical forcing – eccentricity, obliquity and precession. Other possible mechanisms causing rapid climatic changes during glacial periods include the cold Heinrich iceberg melting events (7–13 ka intervals) and the warm Dansgaard–Oeschger cycles (1–3 ka interval, 1500 year duration). Available data therefore indicates that, over timescales of $c.10^5$ years, stability of landform controls is rare and stable process regimes often last for less than 5000 years. This is in the range of known relaxation times for many landform systems.

Over a 10^4 year period, it is possible for temperatures to change by $c.5$–$10°$ C and for there to be recognizable cool-wet and warm-dry periods. For example, it is now established that, over the Holocene, although the transition from glacial to interglacial conditions is largely complete, there have been several major fluctuations. These include the cooling of 8110–4790 BP, the cool event at $c.3500$ BP and the move from the warm, dry sub-Boreal to the cooler, wetter sub-Atlantic of

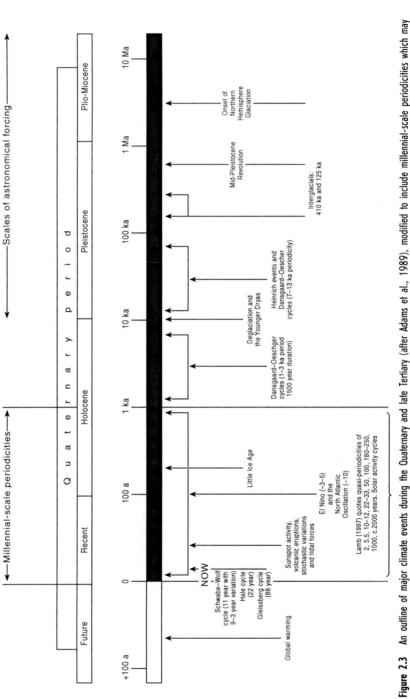

Figure 2.3 An outline of major climate events during the Quaternary and late Tertiary (after Adams et al., 1989), modified to include millennial-scale periodicities which may appear in the 1000 year geomorphological-ecological record. The timescale is logarithmic.

*c.*2500 BP. These have also found expression in significant changes in geomorphological rates of activity.

Over the 10^3 year period of the present discussion, the secular climatic cycles, on centennial and decade scales, that are recorded by most palaeoclimate indicators are more important. These major climatic trends only show as trends on the millennial scale. They may be due to solar activity variation related to sunspot cycles, North Atlantic Oscillation (NAO) and El Niño Southern Oscillation (ENSO) events, volcanic eruptions and stochastic variations (Barber et al., 1994; Roberts, 1998; Waple, 1999). Serious perturbations on the millennial scale can also be caused by natural fires and disease pathogens (e.g. Dutch elm disease), but the geomorphological effects of these on the British landscape over the last 1000 years is unknown. Lamb (1997) also quotes quasi-periodicities, which have been identified from spectral analysis of climatic records of 2, 5.5, 10–12, 22–23, 50, 100, 180–250, 1000 and *c.*2000 years, and suggests energy output from the Sun and variations in tidal forces as likely causes (figure 2.4).

Over the last millennium the British landscape has suffered two important natural perturbations on the 10^2 year scale, the so-called Little Optimum (or Medieval Warm Period) warming of *c.* AD 700–1200 (Trevor-Roper, 1965; Lamb, 1997) and the Little Ice Age (widest definition AD 1420–1850, with a period of amelioration from AD 1500 to 1550). There is quite strong coincidental evidence that these are related to the 'Grand Maximum' and the two great minima of sunspot activity, the Spörer and the Maunder Minima (Adams et al., 1999). The general agreement amongst historians of climate (figure 2.5) is that:

- There was a warm period that perhaps lasted several centuries in the early part of the millennium.
- There was a decline towards wet and cold conditions in the 14–15th centuries, a warm period in *c.* AD 1500–1550 and further cold conditions from that time perhaps until 1850.
- Variable modern conditions with overall warming occurred towards the end of the millennium.
- The onset of each state appears to have been remarkably rapid and there was considerable variability within each phase. Single very warm or cold years occurred against the overall trend.
- Wind directions cycled between strong westerly dominance of greater than 100 days per year and steady east winds during times of anticyclone dominance over the North Atlantic.

From a geomorphological point of view, these conditions seems to have led to the following:

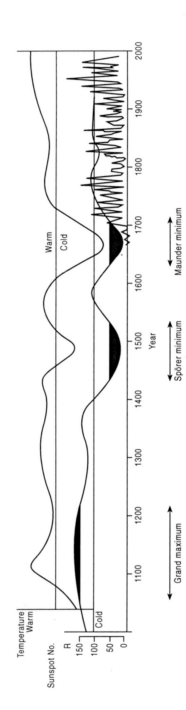

Figure 2.4 The sunspot number for the last millennium, drawn to emphasize the 'grand maximum' and the Sporer and Maunder Minima. Data are generalized before 1700. Temperatures should be regarded as a general cartoon (simplified from US National Geophysical Data Center (in Waple, 1999); Eddy, 1976; Withbroe and Kalkofen, 1994).

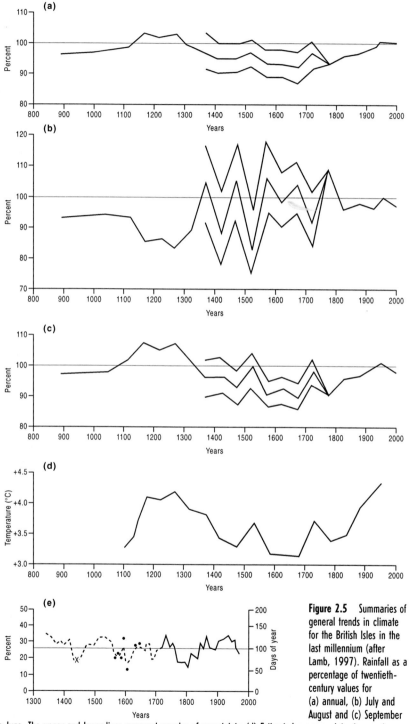

Figure 2.5 Summaries of general trends in climate for the British Isles in the last millennium (after Lamb, 1997). Rainfall as a percentage of twentieth-century values for (a) annual, (b) July and August and (c) September to June. The upper and lower lines represent margins of uncertainty. (d) Estimated average winter temperatures. (e) The frequency of southwesterly winds in England, 1340–1978.

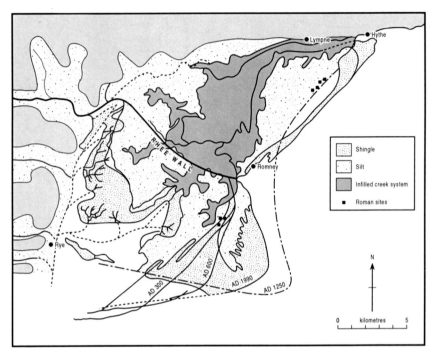

Figure 2.6 The evolution of Dungeness, AD 300–1990 (after Cunliffe, 1980). Note that geomorphological interpretation suggests that the spits and bars grew from the south-west. The change in AD 1250–1990 might also suggest erosion and deposition under the influence of an easterly wind. The shaded creek area indicates that much infilling of inlets has taken place due to sea-level rise and sediment delivery following agriculture.

- There was a slight increase in sea level in the early part of the millennium, perhaps > 1 m over Roman levels, followed by a slight regression over the middle period and a renewed rise to the present (figure 2.2(a)). The slightly higher sea levels that occurred at the beginning of the millennium were very significant on the East Coast. A shallow inlet of creeks extended towards Norwich and the Fenland consisted of channels and islands, including the isolated Isle of Ely. In Dorset and East Devon, wide salt marshes extended up river behind the barrier beaches, and at Dungeness the landforms included wide recurved spits and large tidal lagoons, all of which are now part of the foreland (figure 2.6). Large inlets, which were transgressed by the sea in the early Middle Ages, in East Anglia, the Fens and the Wash, became further infilled in the latter part of the period.
- There was a rapid retreat of soft rock cliffs due to the higher sea

level of AD 1100–1200, and again in the modern period as the sea level continues to rise.

- The paraglacial and para-periglacial supply of sediment to many beaches became exhausted by sea-level rise in the late Holocene. The glacial sediments in the cliffs were rapidly eroded and are now largely exhausted. On the East Coast hundreds of metres of land and many villages have been lost throughout the millennium as these readjustments progressed. There, the supply of material is so large that this adjustment is still proceeding.
- The available sediments were reworked by beach rollover and breakdown into headlands and bays, and deposited into major sediment stores such as Spurn Head, Scolt Head, Orford Ness, Dungeness and Chesil Beach. Many of these areas are suffering cannibalization today, as the complex response mechanisms return to their temporary deposits to supply their current excess sediment transport capacity.
- The variation of the wind directions with the pressure differences across the Atlantic, allied to a high supply of material from the relict sources and the eroding cliffs, must have affected the distribution of the accumulating coastal masses as well as their form. Strong easterly winds for the very long anticyclone periods of the Little Ice Age, for example, must have affected the planform and the pebble size patterns on beaches such as Chesil and Dungeness, which we are accustomed to interpret in terms of a westerly drift. This has never been properly evaluated or modelled, however, and a clear statement cannot be made here.
- Freezing of rivers and the sea, as well as snow lying for long periods in the Little Ice Age, are quite well known. For example, Lamb (1997) quotes a mean snow lay of 20–30 days, against 2–10 days at present, with extremes of 60–70 days for 1862–3, 80 days for 1783–4 and 102 days for 1657–8, in Hertfordshire. The Thames is known to have frozen in winter 20–22 times between 1564 and 1814, and 11 times in the sixteenth century. This is partly because the lower Thames was fresher than now, but the data are still a strong indicator of harsh conditions. There are records of permanent snow patches on Cairngorm and frozen ground to a depth of more than 1 m in Somerset. These all suggest that mechanical weathering, spring thaw, flooding, vicious erosion on clay cliffs and shore platforms due to heavy frost and freezing must all have been enhanced at this time.

There is a clear picture that 'natural' erosion and sedimentation during the early Holocene were related to the adjustment of the system to the deglaciated landscape by rivers of slightly greater power than at present. Sediment flux is also thought to have decreased as available paraglacial sediment was removed, gradients were lowered and revegetation took place (Hooke et al., 1990). Relaxation seems to have become complete by the middle Holocene, although episodic erosion still occurred from the middle to the late Holocene. It might therefore be expected that there would be clear evidence of the effect of environmental change on natural geomorphological process rates during the late Holocene. The sedimentation records in rivers and lakes and on footslope fans is, however, disappointing (Harvey et al., 1981; Richards, 1981; Burrin, 1985; Robertson-Rintoul, 1986; Harvey and Renwick, 1987; Brazier et al., 1988; Hooke et al., 1990). There is little clear evidence of changes in sediment yield or of increased erosion due to natural causes. Perhaps the resolution of our studies is not yet good enough clearly to relate climatic and hydrological causes to the more subtle fluvial changes experienced during the historical second millennium AD, or perhaps we have not looked carefully enough at the landscape indicators and sediment record.

Our understanding of the mass movement record is much better. A number of studies (Starkel, 1966; Hutchinson and Gostelow, 1976; Pitts, 1981; Jones and Keen, 1993; Brunsden and Ibsen, 1994) and compilations (Jones and Lee, 1994; Brunsden and Ibsen, 1997) show that landslides are a sensitive indicator of climatic changes and coastal erosion. The main periods of activity in Britain in the Holocene were the early Atlantic (7500–6000 BP), the sub-Boreal (5500–3000 BP) – a period heavily affected by coastal landslide records due to sea level reaching the preglacial shoreline – and the early sub-Atlantic (2500–2000 BP). In the millennium of interest, the Little Ice Age (AD 1550–1850) and the last 50 years show pronounced increases in activity (table 2.1). The later record may be due to better reporting, but does appear in records of all types and may be a true effect.

2.5.2 Human causes

Perhaps the main reason why the record of the natural causes of geomorphological change in the millennium is not clear in Britain is that human activity has overwhelmed the natural signals. This must be treated as the main perturbation to the stability of the geomorphological system during the last 1000 years. As Roberts (1998) states, 'What makes the last five centuries so distinctive is, of course, the quantum leap in human impact on nature that has occurred during this time.' It is not possible, or necessary, to describe the full range of human activity or the geomorphological effects. These have been brilliantly summarized by Goudie (1977, 1995), Simmons (1996), Roberts (1998) and many others. To advance this discussion, however, it is helpful to note that the main changes that affected the geomorphological processes, namely the clearance of the natural vegetation cover for the first time, the creation of the field patterns, ploughing, mining, road building and drainage, which all played their part in massive landscape change, first occurred in many parts of Britain before the last millennium began. Neolithic, Bronze Age, Iron Age, Celtic, Roman, Saxon and Viking all had an effect and, after 5000 BP, it is their activities that make the first accelerated erosion and sedimentation signals in the geomorphological record.

During the first forest removal phase in any region, the geomorphological rates of change were probably at their highest since the paraglacial period. The forest brown earth developed during the early Holocene, which is still visible under some Bronze Age earthworks, and the forest survivals amongst the heathlands reveal how the unprotected areas suffered severe erosion. On the limestone areas, soil removal is complete. Many valley floors show massive accretion of fine material during the mid-Holocene and occasional incision, depending on the complex response controls of each catchment. On the uplands, the episodicity of cut and fill is particularly well marked. Elsewhere, blanket bog growth provided a new environment and new geomorphological processes.

In reconstructing the changes in the present millennium, the Domesday survey of AD 1086 provides a baseline by which to assess the changes achieved and to come. Over southern and eastern England, only 5% was wooded and the area was densely populated. There were great forests, such as the Forest of Dean, the Caledonian pine woods and the Weald, but over England as a whole only 15% was forested and settlements and organized field systems dominated the rest. In upland Britain and Scotland, some land clearance took place for the first time after AD 1000 particularly as the upper limit of agriculture increased in elevation. There were important erosion effects in the still forested clay lowlands, caused by new forest removal during the expansion and reorganization of agricultural activity of the seventeenth, eighteenth and nineteenth centuries. It is therefore difficult to assess the exact effect of human impacts during the millennium, because the rates have changed in response to the detail of land-use change in each catchment. The sediment yields are a mixture of new clearance, renewed clearance, land use change and urbanization, and these are difficult to distinguish either in the sedimentary record or in the landforms. The main points are as follows:

Table 2.1 The correlation of landslide activity with periods of climatic deterioration during the last millennium (after Ibsen and Brunsden, 1994)

Period zone name	Approximate dates (AD)	Climatic conditions	Inland landslide activity in Britain (Jones and Lee, 1994)	Landslide activity along the south coast of Britain (Ibsen and Brunsden, 1993)
Post-Glacial Modern	1980–2000	Warming trend	Increase in landslide activity often as a consequence of human activity (e.g. coal mining)	Examples: 1999 – Beachy Head; 1986 – Black Ven, Dorset; 1988 – Luccombe, Isle of Wight
	1950–80	Temperature and rainfall variability; general cooling	East Pentwyn and Aberfan, South Wales; road construction – A625 Mam Tor, Derbyshire; M5 Walton's Wood; A21 Sevenoaks; M4 Burderop and Hodson	Examples: 1951 – Luccombe, Isle of Wight; 1952 – Blackgang, Isle of Wight; 1954 – Atherfield, Isle of Wight; 1961 – Beachy Head and Humble Point; 1968 – Budleigh Salterton, Devon; 1968, 1969 and 1978 – Blackgang, Isle of Wight
	1920–50	Warmer period, increase in rainfall and prevailing westerly winds, decrease in snow and frost	Housing development – Hedgemead landslide, Bath; Bury Hill, West Midlands; Brierly Hill, West Midlands; Gypsy Hill, London. Debris flow activity in Scotland	Examples: 1923 – Ventnor, Isle of Wight; 1925 – Bouldner and Ventor Undercliff; 1928 – Blackgang and Lyme Regis; 1932 – St Margaret's Bay; 1935 – Ventnor; 1937, 1940 – Folkestone Warren; 1942 Stonebarrow, Dorset

Period	Date	Climatic characteristics	Landslide activity	Examples
	1850–1920	Climatic warming beginning to occur; rapid industrialization and urban development		Examples: 1858 – Isle of Portland; 1862 – Lyme Regis; 1877 – Folkestone Warren and Stonebarrow; 1886 – Folkestone Warren; 1893 – Sandgate; 1896 – Folkestone Warren; 1905 St Margaret's Bay; 1908 – Black Ven; 1912 – Abbot's Cliff; 1915 – Folkestone Warren
'Little Ice Age'	1550–1850	Appreciable climatic deterioration; cold winters and wet summers, increased frequency of storms and floods	Large number of landslides reported. Examples: 1571 – Wonder at Marcle; 1755 – Whitestone Cliff, North York Moors; 1773 – The Birches, Ironbridge Gorge; 1774 - Hawkey Slip, Selborne, Hampshire; 1790 – Beacon Hill, Bath	Large number of landslides reported. Examples: 1592 – Golden Cap; sixteenth century – Chapel Rock; 1605 – Golden Cap; 1639 – Folkestone Warren; 1665 – Isle of Portland; 1689 – Humble Point; 1724 – Golden Cap; 1765 – Humble Point; 1790 – Downlands; 1792 – Isle of Portland; 1799 – Rocken End; 1810, 1818 – The Landslip, Isle of Wight; 1810 – Dover; 1839 – Bindon, Folkestone Warren; 1840 – Whitlands; 1843 – Beachy Head; 1847 – Shakespeare's Cliff
'Medieval Warm Period'	900–1300	Secondary climatic optimum		Eleventh century – St Catherine's Point, Isle of Wight, Haven Cliffs, Lyme Regis

- The warming of the climate during the Little Optimum of AD 700–1300 was a period of pronounced land use change and enhanced sediment yield, which shows in the pollen record, if not always in the sediment sequences.
- Geomorphological rates of activity did not appear to reach early or mid-Holocene levels during the millennium except in newly cleared areas. Renewed forest clearance took place on the remaining lowlands, tillage of new soils was extended to higher altitudes – to 400 m on Dartmoor and 320 m in Northumberland, for example – and there was very heavy grazing by sheep, widespread vineyard establishment and considerable technological, population and urban growth. These changes must have caused high local sediment yields. General geomorphological studies suggest that newly cleared or fired land has up to 1000 times the sediment yield of naturally forested areas (Dearing and Foster, 1993; Goudie, 1995), but such rates would only have been reached in the newly cleared areas and not those previously cleared in earlier millennia. In the urban areas initial rapid and heavy sediment yields gave way to rapid run-off without non-urban sediment.
- Extensive river diversion and straightening (figure 2.7) provided sediment for a new phase of coastal salt marsh accumulation in nearly all estuaries and coastal inlets, where reclamation of grassland became a strong stabilizing influence. The Wash was heavily affected due to the vulnerability of the soils of the Fens to erosion (figure 2.8(a)).
- The combined effects of climatic change and human activity had a clear impact on the rivers. In addition to construction and drainage changes, there were sedimentation phases related to the type of activity in each catchment, such as mining and metal production (figure 2.8(b)). There are also minor-scale examples of river channel changes, both of braided reaches (Werritty and Ferguson, 1980) and meandering rivers. There was some aggradation of floodplains at this time, which produced low terraces when followed by Little Ice Age incision (Hooke et al., 1990). Such changes are well documented by the Ordnance Survey for the recent historical period.
- Pond construction, peat digging on a massive scale (2 500 000 m^3 of peat removed to form the Norfolk Broads), moats and marl pits disfigured the land surface but acted as local sediment traps. Meres and small lakes showed accelerated accumulation in medieval times, especially where new ditches drained the land (Edwards and Rowntree, 1980).
- The higher tree lines, high tillage and grazing in the Medieval Warm

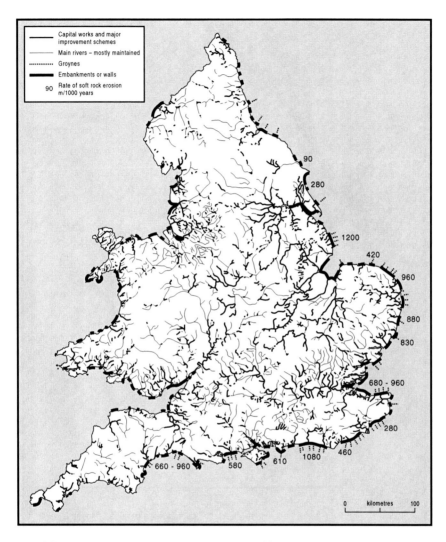

Figure 2.7 Human regulation of rivers and the coast in the last 300 years (after Brookes et al., 1983; Goudie and Brunsden, 1994).

Period had a greater effect on the stability of slopes in the uplands, and several authors have shown that the frequency of debris flows, soil creep and gully erosion were all related to land use changes. A wave of soil erosion, gully development and debris cone deposition probably followed the Scandinavian introduction of sheep farming to the Howgill Fells in the tenth century (Harvey et al., 1981).

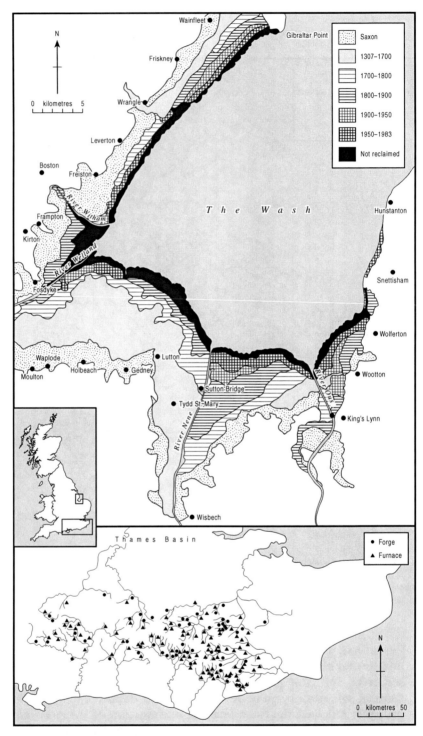

Figure 2.8 Examples of effective land 'forming' human activity during the last millennium. (a) Land reclamation around the Wash (after Doody and Barnett, 1987). (b) Wealden medieval blast furnaces and forges, requiring hammer ponds and forest clearance (after Burrin, 1985).

Human interference and burning destroyed the vegetation cover and caused fluvial incision and development of alluvial fans of reworked sediment in Glen Etive of the Grampian Highlands in AD 1305–1425 (Brazier et al., 1988). Low terraces were also formed by an aggradation–incision phase in Glen Feshie at the beginning of the millennium (Robertson-Rintoul, 1986). There appears to be a consistent pattern of an early (Holocene) and a late (this millennium) phase of erosion and fan deposition in many upland areas.

- Despite this, however, the indications are that the stability of the land surface was quite normal and that only certain sensitive areas were subject to large events. For example, there are only two large landslides reported for this period, at St Catherine's Point, Isle of Wight, and at Haven Cliff, Lyme Regis (Jones and Lee, 1994).

During the Little Ice Age many of these trends were reversed or temporarily halted, as people reacted to the harsh living conditions and retreated from the moorland edges or abandoned land. Once again, although we know a great deal about the history of the landscape throughout the fifteenth to eighteenth centuries, the exact geomorphological changes and rates of activity are far from clear. The same might be said of the seventeenth to eighteenth-century climatic recovery period, and it is only when we examine modern recorded events that we can reconstruct reasonably accurate event series.

The fact that there is less evidence of a strong aggradation phase in estuaries and lowland sediment sinks, and that debris cones and fans of debris of these periods still persist in the uplands, suggests that sediment delivery processes were not always very efficient. There was considerable storage within the catchments, in ponds or in lakes (Edwards and Rowntree, 1980) and there was poor slope–channel coupling. This was perhaps due to the strong growth of floodplains in the paraglacial and mid-Holocene periods, that have acted as a barrier to sediment removal from the foot slopes in the last 1000 years. The more modern development of reservoirs and flood control works on major rivers has enhanced this effect, as have the widespread use of hedges and banked boundaries. The various enclosure movements should perhaps be regarded as one of the most important historical stabilizing forces in the evolution of the landscape. Unfortunately, we do not know the quantities involved. In particular, although human activity may have created as many barriers and sediment traps as it did erosion hotspots, we do not know how much 'millennium' sediment is in mid-system storage.

An indicator of the inefficiency of late-millennium sediment coupling processes is the widespread survival of ridge-and-furrow. These low, linear structures are no more than 1 m high and are only made of topsoil, but in some cases denudation has been unable to remove them, even where they were abandoned as maintained arable field features several centuries ago. Although every new construction or change must have shifted some soil and increased the overall sediment yield, we simply have no idea of the true amounts except by analogy to modern examples (e.g. Walling and Gregory, 1970), which suggests increases of 10–100 times the undisturbed figure. The truth may be that the sediment yields of the present system are derived from the erosion of the channel banks and bed, the headwater areas where there are no floodplains, the drainage ditches of arable fields and disturbance during urban development. In some areas not directly coupled to the channels, current sediment yield may merely be reworking early to mid-Holocene stored sediment as part of the fluvial complex response system. Certainly, there are several records (e.g. Hooke et al., 1990) which show that the last 1000 years have been characterized on lowland rivers by an early deposition of a 'Low Terrace' followed by shallow and continuing incision. A generalized conclusion might be that erosion is very localized, and it may make no sense to talk of overall denudation or lowering rates in terms of the area of a landscape or a catchment.

2.6   Formative Events

When we consider the last part of the millennium it is helpful to consider proposition 6, which states that '*landforms are produced by specific formative process events*', and proposition 7, '*that new landforms are produced when the normal behaviour of a system is overturned by geocatastrophes*'.

On the decade timescale, attention focuses on short-term variability of the system, cycles and trends, episodes (such as a sequence of wet years), extreme events and regular, repetitive occurrences or behaviour. Short cycles, which appear to explain part of the climatic variability, are the Gleissberg cycle (88 years), the Hale cycle (22 years) and the Schwabe–Wolf cycle (11 years, with 9–13 year variability). These have certainly been effective over the last 500 years and, when translated into temperature, pressure distribution, wind direction and precipitation (figure 2.9), may well explain some of the geomorphological variability shown in the historical flood frequency, mass movement and beach drift direction records.

The main features of the millennium record seem to be the following:

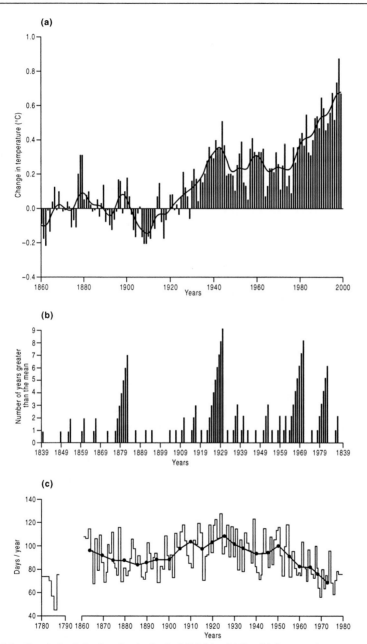

Figure 2.9   Recent climatic trends and cycles for the British Isles. (a) The global mean surface air temperature, 1860–1999, with a superimposed running average (after Hadley Centre, 1999). (b) Cumulating moisture balance greater than the mean for the Isle of Wight, 1839–1987 (after Brunsden and Ibsen, 1997). (c) The number of days each year with westerly winds blowing over the British Isles, 1781–5 and 1861–1979. The solid line represents a 10 year running mean, plotted at five year intervals.

- The underlying warming–cooling, drying–wetting perturbations of the millennium scale show as the long phase trends of the decade scales. They are accompanied by short-period cycles, such as the NAO, which are reflected in the known geomorphological histories (figure 2.10).
- For example, the migration of blown sand has been a hazard in Scotland, Wales and the northern countries of the Atlantic throughout the millennium. In 1316 at Kenfig, South Wales, the medieval port was closed by dunes after many years of advance. Between 1385 and 1400 at Morfa Harlech, dunes overwhelmed the port. At Culbin Sands, 60 km² of farmland and nine farms were lost, and there was severe wind erosion in Breckland. At the Sands of Forvie, advancing dunes 30 m high finally moved over the town in the great storm of 1413. Some of the major hazard events are summarized in chapter 10.
- Along the South Coast of England, the mass movement databases (Brunsden and Ibsen, 1994) reveal a clear association of wet year cycles and landslide years. It is also likely that the cliff retreat rate varied in a cyclical manner (figure 2.10(b)).
- The historical reconstructions of Lamb (1997) show that single high-magnitude events were very important throughout the millennium, and the measured record shows that this continues today. Noteworthy early in the millennium (AD 1240–1362) was the occurrence of increased storms, severe winds and sea floods all around the North Sea (figure 2.10(e)). The loss of the ports of Ravenspur and Dunwich are famous examples of geomorphological damage, although these must have been due to progressive coastal retreat as well as the definitive storm. Lamb cites the years of 1421, 1446, 1570 and 1634 as bad years all around the North Sea. Over 100 000 lives were lost, in this whole area, on two occasions. There is increasing evidence that recent years are again beginning to be dominated by large events.
- The mass movement record confirms the idea that single events can be formative in the sense that they create, in one attempt, a landform that survives for a long time as a diagnostic element of the landscape assemblage. Examples range across the whole country, from the Wonder at Marcle at the Woolhope Dome of 1571 to the failure of the Bindon landslide of Christmas Eve 1839 in East Devon (table 2.1). The events were complete in days; the forms are still there for all to see (Frenzle et al., 1997). The collapse of Mam Tor in the Peak District in 8000 BP, which is still an important landform, suggests that formative events can have very long lifetimes indeed, although in this case the features have been 'freshened up' on more

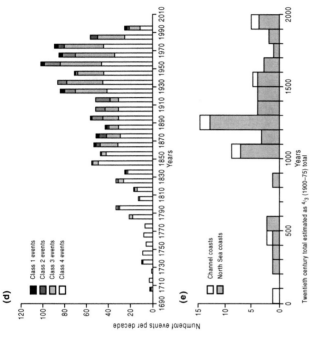

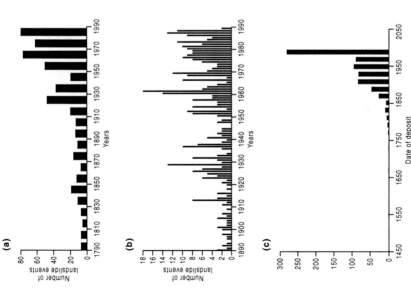

Figure 2.10 A geomorphological event record, which shows an apparent rise in the number of events towards the present. Care must be exercised when using these data, since the effect of reporting efficiency is unknown. Where the measured record (e.g. rainfall) agrees with an event record (e.g. floods, landslides) the record can be used with more confidence. (a) The number of landslide events per decade, 1790–1980. (b) The number of events per year, 1890–1989, on the South Coast of England between Straight Point and St Margaret's Bay (after Ibsen, 1994). (c) The frequency distribution of debris flow deposits at sites in the Scottish Highlands, obtained from lichenometry (after Innes, 1983). (d) The pattern of damaging flooding and coastal erosion events by decade, 1700–1993. Class 1 Moderate to Class 4 Major (after Lee, 1995). (e) The number of severe sea floods per century (after Lamb, 1997).

than one occasion. Events formed in this millennium may be part of the landscape ten millennia from now!

- Examples of important events in the last century include sudden channel changes (Hooke et al., 1990); the drowning of the Broads by a cataclysmic surge of the North Sea in 1287 (Rackham, 1987) and again on the East Coast in 1953; the production of a fan delta at Charmouth, Dorset by a flood in 1972; and the landslides, channel changes, transport of enormous boulders, alluvial fan and damage of the Lynton and Lynmouth storm of 1952 and a previous event in 1796 (Anderson and Calver, 1980). Human activity also generates big events: Aberfan in 1966; solution collapses at Ripon, Yorkshire, several times since 1800; or the landslide of Lyme Regis in 1962. The list could, of course, be extended.

The point is that, throughout the landscape, forms that have been produced in a very short time may persist for a long time, or the consequences may affect future change for centuries. Human events, such as the building of a hilltop fort by the Iron Age peoples (still a landform after 2000 years of degradation), the construction of a reservoir or the M1, all illustrate the point. Although it is a common view that most geomorphological work is done by the regular repetition and cumulative effect of everyday events (and this may still be true of human activity), we would do well to remind ourselves that some persistent forms may be produced in a very short time. This is increasingly true of this millennium, because human beings have the power completely to eradicate natural systems or permanently change their operation. This may prove to be the most distinctive characteristic of the period and the most important warning for the future.

Proposition 8 states that '*When a perturbing displacement exceeds the resistance of the system, the system will react and relax towards a new stable state which will be expressed by a new characteristic form.*' This subject is of importance because it concerns the ability of a system to recover from a disturbance and is therefore a consideration in land management.

The evidence suggests that the process systems of the millennium have been benign. The rates of activity are have been low even where they are accelerated by human activity and current relaxation times are long. Despite this, on beaches, in channels or in the tracks of dynamic landslides, landforms can quickly reach a steady state or dynamic equilibrium with the current sediment flux. Some examples are found in the work of Slaymaker (1972) for the River Wye, Brunsden and Chandler (1997) for the Black Ven coastal mudslide and Brunsden and Jones (1976) for the rotational landslide complex of Stonebarrow Hill, Dorset.

Hutchinson and Gostelow (1976) have shown that the relaxation time for a hill slope to become stable against landsliding on London Clay may be as long as 10^4 years, but that the form of each failure episode is visible long after episodic failures have ceased. Data for landslide scars in soft sandstone cliffs show that recovery – in the sense of the creation of a fully vegetated scar with no bare ground and low erosion – can take as little as 10^2 years if there has been no further disturbance. The development of a new soil and mature vegetation on the talus may take 10^2-10^3 years but, as discussed earlier, the removal of the landslide scar or the large landslide blocks as a landform is on the 10^3-10^5 year timescale. Landslides which occurred early in the millennium are still very much with us!

Although the signs of a large event may remain visible on hill slopes for long periods, on rivers it is often difficult to identify the erosion effects of any of the big storms which have occurred in the last 100 years. Re-vegetation and human repair can be very efficient. The achievements of the massive 1968 storm on Mendip, which ripped out the floor of Cheddar Gorge, filled some of the caves and the ponds with sediment and stripped the side valley floors, are now almost unidentifiable, even by the geomorphologists of the 'Floods Patrol' who carried out the post-storm surveys. Channels may recover their former channel shape and sediment yield after a storm in perhaps 10^1-10^2 years. This may also be true of the shallow landslide scars caused by a flood if the channel–hillslope coupling is good, such as in the case of the Exmoor storm of 1952 (Anderson and Carver, 1980).

Paradoxically, many floodplains still carry the marks of channels created in storms early in the millennium and the terraces and incised channels produced earlier in the Holocene. The same is true of Neolithic–Iron Age settlement structures and such weak features as ridge-and-furrow. The trenches and furrows may have been partly filled in and the 'sharp' edges rounded off but, amazingly, little has been done to the overall form by current processes, even though ploughing ceased several hundred years ago.

The lesson provided by the evidence of the last millennium is that we should distinguish between the ability of a system to recover to its previous behaviour pattern or to establish a new stable process regime and sediment flux, and the ability of the system to remove evidence of major geometrical change. The former has been easily achieved in 100 years; the latter has not except, where the system is very dynamic.

The main reason for this is expressed by the last two propositions, which concern the sensitivity of a landscape to change. Proposition 9 states that, '*There is in landscape a wide spatial variation in the ability of the landforms to change. This is known as the sensitivity to change. Thus landscape sensitivity is a function of the temporal and spatial distributions of the resisting and disturbing forces and is therefore diverse and complex.'* Proposition 10 suggests that, *'The ability of a landscape to resist impulses of change tends to increase with time.'*

The impulses of change on any landscape are so numerous and varied that it is surprising that we do not observe constant dynamic behaviour. Instead, except in the most labile of systems such as coastal cliffs, we usually record a very complex and episodic response. This is certainly true of the British landscape over the last 1000 years. So far, however, we have only considered the driving forces for change – tectonic, seismic, eustatic, climatic and human. To balance the 'sensitivity' equation it is necessary to consider the magnitude and distribution of the barriers to change.

The main barrier on the 1000 year timescale is the morphological resistance of the system. This is very strong, due to the history of an effective, preceding periglacial regime and successful paraglacial adjustment to the debris supply in the uplands. The rate of activity of the modern anthropogenic, accelerated erosion processes has been preconditioned by the Neolithic–Norman forest clearance and agricultural activity. In some areas these activities have already done much of the possible work, and a high percentage of the land surface has been sterilized under urban and industrial structures, so that it is no longer able to change by natural processes (figure 2.11). Very large areas of the system have very flat slopes, little mobilizable potential energy and slow rates of activity. Even the uplands have surfaces of low relief due to their plateau form and ancient etch plain origin. Much of the landscape is therefore both benign and is either readily brought back to a fully relaxed state after the application of a perturbation or does not react at all.

The system possesses considerable strength resistance, because the exposed rock types in upland Britain are strong, and even in the less resistant Mesozoic rocks the contemporary progressive weathering processes are slow. The weathering rates generally lag behind the removal processes, so that the systems are supply limited. There is also high locational resistance, in that contemporary sea-level changes in much of Britain, except Scotland, are diminishing relief and drowning valleys.

There is increased transmission resistance and a greater ability to transmit energy, because the pathway density of ditches and streams has been improved, but this is counteracted by the control and run-off by dredging, reservoirs, riparian strips for erosion control, ponds, hedges and lack of drainage ditch maintenance. These controls, added to the effect of wide Holocene floodplains, means that hillslope–channel

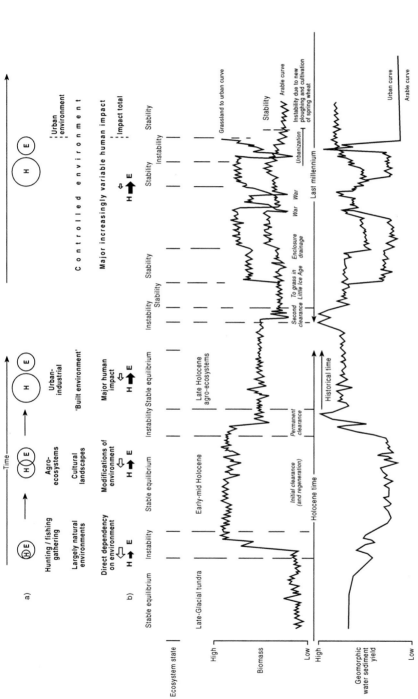

Figure 2.11 A cartoon of the changing relationship between humans (H) and the natural environment (E), as presented by Roberts (1998) for the Holocene, to indicate the nature of interaction and relative impact, extended for the last millennium, noting some historical changes of land clearance, war changes in land use, enclosure and change in urban areas.

coupling is weak, lengthy and contorted. Erosion activity is therefore concentrated in the channels and headwaters, or in areas of steep gradient. The shock absorbers of the system, such as mobile river-bed sediments, floodplains, reservoirs and high storage capacity soils and agricultural structures (enclosed fields), are numerous and effective. There is very little overland flow in this system. Only the beaches, which are beginning to diminish as the paraglacial supply runs out, coastal cliffs, and certain very unusual sensitive areas, such as the exposed fragile soils of the Fens or the headwaters of steep streams, can be regarded as areas that are sensitive to change.

The lessons of the millennium are that the British landscape may be regarded as relatively insensitive. It has low relief and elevation energy, propagates events slowly, has large storage, and low concentrations of energy on the erosion axes, and has readily absorbs perturbations such as the Little Ice Age. Following massive human aggression and ecological transformation such as renewed land clearance, land-use change and wartime effects, sudden pulses of geomorphological work took place, but the inherent stability of the barriers to change meant that geomorphological stability was quickly restored. There is therefore an overall persistence of relief and pattern, a stagnancy of development and a palimpsest of forms.

Despite this, there is a considerable sting in the tail. Although first-time events such as slope failures are rare (inland) and cataclysmic events (typhoons, large earthquakes and major floods) are infrequent, the high-energy absorption capacity and long lifetimes of landforms means that there is an accumulation of relict features and storage of sediment. These can be reactivated by both slow acting but progressive forces, and particularly by human activity. This means that the landscape over the millennium and in the future is capable of dramatic and large-scale change if we inadvertently fail to read the signs. Many of the millennium's major slope failures in Britain have been due to human disturbance of relict forms. It is certain that this will be a major feature of the next millennium, with or without the effects of global warming.

REFERENCES

Adams, J., Maslin, M. and Thomas, E. 1999: Sudden climatic transitions during the Quaternary. *Progress in Physical Geography*, 23, 1–36.
Anderson, M. G. and Calver, A. 1980: Channel plan changes following large floods. In R. A. Cullingford, D. A. Davidson and J. Lewin (eds) *Timescales in Geomorphology*. Chichester: Wiley, 43–52.
Barber, K. E., Chambers, F. M., Maddy, D., Stoneman, R. and Brew, J. S.

1994: A sensitive high resolution record of late Holocene climatic change from a raised bog in northern England. *The Holocene*, 4, 198–205.

Brazier, V., Whittington, G. and Ballantyne, C. K. 1988: Holocene debris cone evolution in Glen Etive, western Grampian Highlands, Scotland. *Earth Surface Processes and Landforms*, 13, 525–31.

Brooks, S. M., Richards, K. S. and Anderson, M. G. 1993: Modelling the sensitivity of slope stability to mass movement during the Holocene, In D. S. G. Thomas and R. J. Allison (eds), *Landscape Sensitivity*. Chichester: Wiley, 15–27.

Brookes, A., Gregory, K. J. and Dawson, F. H. 1983: An assessment of river channelisation in England and Wales. *Science of the Total Environment*, 27, 97–112.

Brunsden, D. 1990: Tablets of stone: toward the ten commandments of geomorphology. *Zeitschrift für Geomorphologie, Supplementband* 79, 1–37.

Brunsden, D. and Chandler, J. H. 1997: Development of an episodic landform change model based upon the Black Ven mudslide, 1946–1995. In M. J. Anderson and S. M. Brooks (eds), *Advances in Hillslope Processes*, Vol. 2. Chichester: Wiley, 869–96.

Brunsden, D. and Ibsen, M.-L. 1994: The nature of the European archive of historical landslide data, with specific reference to the United Kingdom. In R. Casale et al. (eds), *Temporal Occurrence and Forecasting of Landslides in the European Community*. Final report, EPOCH Programme, European Commission, Science Research Development, Ref. EUR 15805 EN, 1, 21–70.

Brunsden, D. and Ibsen, M.-L. 1997: The temporal occurrence and forecasting of landslides in the European Community: summary of relevant results of the European Community EPOCH Programme. In B. Frenzle, J. A. Matthews, D. Brunsden, B. Glaser and M. M. Weiss (eds), *Rapid Mass Movement as a Source of Climatic Evidence for the Holocene*. Stuttgart: Gustav Fischer Verlag.

Brunsden, D. and Jones, D. K. C. 1976: The evolution of landslide slopes in Dorset. *Philosophical Transactions of the Royal Society, London*, A283, 605–31.

Brunsden, D. and Thornes, J. B. 1979: Landscape sensitivity and change. *Transactions of the Institute of British Geographers*, NS, 4, 303–484.

Burrin, P. 1985: Holocene alluviation in southeast England and some implications for palaeohydrological studies. *Earth Surface Processes and Landforms*, 10, 257–72.

Carter, R. W. G. 1988: *Coastal Environments: an Introduction to the Physical, Ecological and Cultural Systems of Coastlines*. London: Academic Press.

Clayton, K. M. 1997: Review of S. B. McCann and D. C. Ford (eds) *Geomorphologie sans Frontiers*. *Earth Surface Processes and Landforms*, 22, 1085.

Cooper, A. H. and Waltham, A. C. 1999: Photographic feature: subsidence caused by gypsum dissolution at Ripon, North Yorkshire. *Quarterly Journal of Engineering Geology*, 32, 305–10.

Cunliffe, B. W. 1980. The evolution of Romney Marsh: a preliminary statement. In F. H. Thompson (ed.), *Archaeology and Coastal Change*. London: Society of Antiquaries, 37–55.

Dearing, J. A. and Foster, I. D. L. 1993: Lake sediments and geomorphological processes: some thoughts. In J. McManus and R. W. Duck (eds), *Geomorphology and Sedimentology of Lakes and Reservoirs*. Chichester: Wiley, 5–14.

Doody, P. and Barnett, B. (eds) 1987: *The Wash and its Environment*. Peterborough: Nature Conservancy Council.

Eddy, J. 1976: The Maunder Minimum. *Science*, 192, 1189–202.

Edwards, K. J. and Rowntree, K. M. 1980: Radiocarbon and palaeoenvironmental evidence for changing rates of erosion at a Flandrian stage site in Scotland. In R. A. Cullingford, D. A. Davidson and J. Lewin (eds), *Timescales in Geomorphology*. Chichester: Wiley, 207–23.

Frenzle, B., Matthews, J. A., Brunsden, D., Glaser, B. and Weiss, M. M. (eds), 1997: *Rapid Mass Movement as a Source of Climatic Evidence for the Holocene*. Stuttgart: Gustav Fischer Verlag.

Goudie, A. S. 1977: *Environmental Change*. Oxford: Clarendon Press (2nd edn, 1983).

Goudie, A. S. 1995: *The Changing Earth: Rates of Geomorphological Processes*. Oxford: Blackwell.

Goudie, A. S. and Brunsden, D. 1994: *The Environment of the British Isles*. Oxford: Oxford University Press,.

Gray, J. M. 1992: Raised shorelines. In M. J. C. Walker, J. M. Gray and J. Lowe (eds), *The South West Scottish Highlands*. Field Guide, Quaternary Research Association, 81–4.

Greensmith, J. T. and Tucker, E. V. 1973: Holocene transgressions and regressions on the Essex coast, outer Thames estuary. *Geologie en Mijnbouw*, 52, 193–202.

Hadley Centre 1999: *Climatic Change Scenarios for the United Kingdom*. Bracknell: Hadley Centre.

Harvey, A. M. and Renwick, W. H. 1987: Holocene alluvial fan and terrace formation in the Bowland Fells, Northwest England. *Earth Surface Processes and Landforms*, 12, 249–57.

Harvey, A. M., Oldfield, F., Baron, A. F. and Pearson, G. W. 1981: Dating of post-glacial landforms in the central Howgills. *Earth Surface Processes and Landforms*, 5, 401–12.

Hooke, J. M., Harvey, A. M., Miller, S. Y. and Redmond, C. E. 1990: The chronology and stratigraphy of the alluvial terraces of the River Dane Valley, Cheshire. *Earth Surface Processes and Landforms*, 15, 717–38.

Hutchinson, J. N. and Gostelow, T. P. 1976: The development of an abandoned cliff in London Clay at Hadleigh, Essex. *Philosophical Transactions of the Royal Society, London*, A283, 557–604.

Ibsen, M.-L. 1994: Evaluation of the temporal distribution of landslide events along the south coast of Britain between Straight Point and St Margarets Bay. M.Phil. thesis, University of London, unpublished.

Ibsen, M.-L. and Brunsden, D. 1994: The temporal causes of landslides on the south coast of Great Britain. In R. Casale, R. Fantechi and J.-C. Flageollet (eds), *Temporal Occurrence and Forecasting of Landslides in the European Community*. Final Report (EPOCH), EC Programme, European Commission, EUR 15805, 2, 322–39.

Innes, J. L. 1983: Lichenometric dating of debris-flow deposits in the Scottish Highlands. *Earth Surface Processes and Landforms*, 8, 579–88.

Jones, R. L. and Keen, D. H. 1994: *Pleistocene Environments in the British Isles*. London: Chapman and Hall.

Jones, D. K. C. and Lee, E. M. 1994: *Landsliding in Great Britain: a Review*. London: HMSO.

Lamb, H. H. 1997: *Climate, History and the Modern World*. London: Routledge.

Lee, E. M. 1995: *The Occurrence and Significance of Erosion, Deposition and Flooding in Great Britain*. London: HMSO.

McCann, S. B. and Ford, D. C. (eds) 1996: *Geomorphology sans Frontieres*. Chichester: Wiley.

Pitts, J. 1981: Landslides of the Axmouth–Lyme Regis undercliffs National Nature Reserve, Devon. Unpublished Ph.D. thesis, University of London.

Rackham, O. 1987: *The History of the Countryside. The Full Fascinating Story of British Landscape*. London: Dent.

Richards, K. S. 1981: Evidence of Flandrian valley alluviation in Staindale, North York Moors, *Earth Surface Processes and Landforms*, 6, 183–6.

Roberts, N. 1998: *The Holocene: an Environmental History*. Oxford: Blackwell.

Robertson-Rintoul, M. S. E. 1986: A quantitative soil stratigraphic approach to the correlation and dating of post-glacial river terraces in Glen Feshie, western Cairngorms. *Earth Surface Processes and Landforms*, 11, 605–17.

Schumm, S. A. 1976: Episodic erosion: a modification of the geomorphic cycle, In W. Melhorn and R. Flemal (eds), *Theories of Landform Development*. Binghamton, New York: Allen & Unwin, 69–85.

Shennan, I. 1989: Holocene crustal movements and sea-level changes in Great Britain. *Journal of Quaternary Science*, 4, 77–89.

Simmons, I. G. 1996: *Changing the Face of the Earth: Culture, Environment, History*, 2nd edn. Oxford: Blackwell.

Sissons, J. B. 1992: Shorelines and isostasy in Scotland. In D. E. Smith and A. G. Dawson (eds), *Shorelines and Isostasy*. London: Academic Press, 209–25.

Slaymaker, O. 1972: Patterns of present sub-aerial erosion and landforms in mid-Wales. *Transactions of the Institute of British Geographers*, 55, 47–68.

Starkel, L. 1966: The palaeogeography of mid- and eastern Europe during the last cold stage and west European comparisons. *Philosophical Transactions of the Royal Society, London*, B280, 351–72.

Stoddart, D. R. 1978: Chinese geomorphology: some conclusions. *Geographical Journal*, 144, 205–7.

Trevor-Roper, H. 1965: *The Rise of Christian Europe*. London: Thames and Hudson.

Waple, A. M. 1999: The sun–climate relationship in recent centuries: a review. *Progress in Physical Geography*, 51, 309–28.

Walling, D. E. and Gregory, K. J. 1970: The measurement of the effects of building construction on drainage basin dynamics. *Journal of Hydrology*, 11, 129–44.

Werritty, A. and Ferguson, R. I. 1980: Pattern changes in a Scottish braided river over 1, 30 and 200 years. In R. A. Cullingford, D. A. Davidson and J. Lewin (eds), *Timescales in Geomorphology*. Chichester: Wiley, 53–68.

Withbroe, G. L. and Kalkofen, W. 1994: Solar variability and its terrestrial effects. In J. M. Pap et al. (eds), *The Sun as a Variable Star: Solar and Stellar Irradiance Variations*. Cambridge: Cambridge University Press.

Chapter 3

The Evolution of Hillslope Processes

David K. C. Jones

3.1 Introduction

The last 1000 years represents an unusually awkward time-frame for
geomorphologists, for it extends well beyond the limits of instrumental
records, careful observations and creditable archival information, but
not far enough to be embraced wholeheartedly by Quaternary science.
It is also rendered problematic in the British Isles because of the
dominant imprint of human activity, especially during the last 200
years, for anthropogeomorphology has never been particularly fashion-
able in Britain. This undoubtedly reflects the evolution of geomorphological thinking in the United Kingdom during the twentieth century,
which was dominated first by long-term landform evolution or historical
geomorphology and then by process studies. To many, the role of
human activity was traditionally seen as a contemporary nuisance,
creating anomalous 'artificial' embellishments and conditions which
obliterated evidence and obscured 'natural' process–landform relationships, and it is only during the last two decades that significant attention
has begun to be focused on the role of humans in altering the rate of
process operation and in modifying surface form (Goudie, 1993, 1995).

It follows from the above that the last 1000 years has also tended to
be viewed as representing a somewhat anomalous period of time with
respect to landform development, falling as it does within the human-
transformed upper Holocene. Indeed, growing awareness of the extent,
thickness and variety of accumulating surficial deposits either produced
by human activity (e.g. made ground) or containing detritus of human
origin of use for dating purposes, led to the amusing French division of
the upper Holocene into *Poubellien supérieur (á plastique)* and *Poubellien
inférieur (sans plastique)*, or Upper Dustbinian (with plastic) and Lower

Dustbinian (without plastic) (see Ager, 1973). There is much to commend this idea and its incorporation within a new epoch that could be called the 'Anthropocene'.

Exactly when such an Anthropocene should be seen to replace a truncated Holocene is a matter of debate, depending on the criteria adopted. Some would undoubtedly advocate AD 1000, which roughly equates with the beginning of the industrial cycle of global population growth (Whitmore et al., 1990), while others would prefer *c.* AD 1500, so as to conform with what Roberts (1998) refers to as the culmination of the 'taming of nature' phase and the commencement of 'the modern era'. However, there is likely to be even greater support for *c.* AD 1750, approximately corresponding to the start of the Industrial Revolution, the commencement of anthropogenically enhanced greenhouse gas emissions and the beginning of the 'The Great Climacteric' (Burton and Kates, 1986), an ongoing period which has seen fundamental changes in the interrelations between sociotechnical systems and the physical environmental systems, resulting in profound effects on the latter as testified by Kates et al. (1990, 1): 'Most of the change of the past 300 years has been at the hands of humankind, intentionally or otherwise.' Irrespective of the details or the terminology used, it is abundantly clear that a time-span as short as the last 1000 years involves consideration of surprisingly great changes in environmental conditions due to human activities.

3.2 The Background to Hillslope Evolution Over the Last 1000 Years

Brown (1979, 455) clearly stated that the last 1000 years in Britain 'might have been a period of relative geomorphological stability at the culmination of the post-glacial rise in temperature and under a virtually complete cover of forest' but for human intervention. Traditional ideas of the last millennium as being geomorphologically inactive have to be discarded and replaced by the view that human activity, combined with relatively minor changes in climate, have had significant consequences in terms of landform development. However, to attempt to generalize about influences on hillslope processes and development in Britain over the last 1000 years is difficult for three fairly obvious reasons:

1 The latitudinal extent of Britain, its elevational differences and the variety of geological and terrain conditions all emphasize diversity in terms of evolutionary histories and responses to change.
2 The record of climate change over the last 1000 years remains to be clearly elucidated.

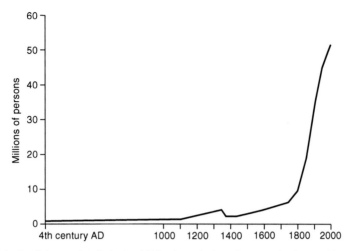

Figure 3.1 Population growth in England and Wales during the last millennium (after Stamp, 1964). The ability to achieve geomorphological work must also be considered to have increased several-fold per head of population during this period.

3   The variations in human impact over space and through time still remain to be clearly established, delimited and quantified.

The last mentioned is clearly the most important. Human impact on the vegetation of Britain is known to extend back at least 8000 years, but was given a significant boost by the arrival of the plough in about 5000 BP. Despite the small population – estimated to be still under 1.5 million at AD 1100 (figure 3.1) – much of the original forest cover was progressively removed relatively quickly to create heath and farmland. As Oliver Rackham (1985, 104) has stated, 'The Romans and their predecessors had reduced well over half of England to farmland and moorland'. At the time of the Domesday survey it is estimated that woodland covered only about 15% of Lowland England and by AD 1350 most of the 'Great Woods' had gone, and woodland had been further reduced to 10% and considerably less in some areas (Rackham, 1985, 1986). Thus Oliver Rackham's fundamental division of Lowland Britain (figure 3.2) into an intimate and intricate 'Ancient Countryside' of irregular woods, hedged fields, winding lanes and scattered hamlets on the one hand, and a less wooded, much more regular 'Planned Countryside' of big fields and villages on the other, reflects a pattern of clearance that very considerably predates the final fashioning achieved by the Enclosure Acts of the 18th and 19th centuries.

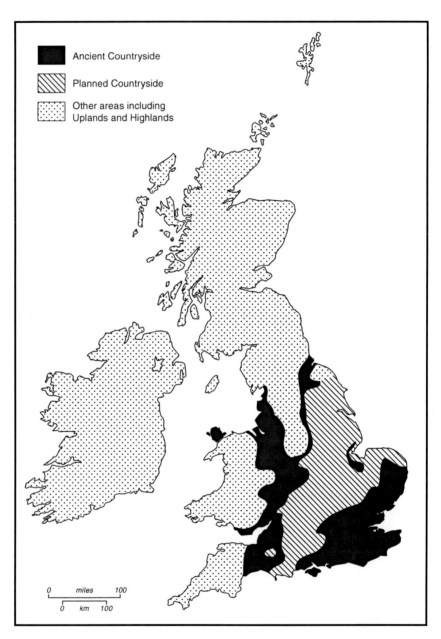

Figure 3.2   Division of the lowland zone of England and Wales into Ancient Countryside and Planned Countryside (based on Rackham, 1986).

There is similar evidence for significant early forest clearance in Scotland and Wales, thereby suggesting three broad conclusions:

1   That patchy clearance led to the occurrence of early phases of soil erosion going back at least 5000 years, as testified by lake and valley sedimentation and the existence of truncated soils.
2   That such episodes may well have been accompanied by increased landslide activity.
3   That by AD 1000 human influences were both widespread and locally significant. Indeed, Price (1983, 196) was moved to state, 'What is certain is that over the period 1000 AD to the present day changes in the environment have been far less important in changing the Scottish landscape than the activities of humans'. What is true for Scotland is undoubtedly true of other parts of Britain.

Despite the mounting evidence for the long-term and profound influence of human activity on slope processes, it is nevertheless important to place these influences in the context of natural changes in environmental conditions over the millennium. The various lines of evidence – observational, textural, sedimentological and biotic (including tree ring studies) – have combined to produce a steady improvement in knowledge of climatic change in Britain over the millennium, which has been summarized in chapter 1. Nevertheless, it has to be recognized that the detail is good only from the present until the 1840s, then increasingly poor back to the commencement of the England and Wales Precipitation (EWP) series in 1766 (1757 for Scotland) and the Central England Temperature (CET) record in 1659, prior to which the evidence is fragmentary. Despite the shortcomings, three distinct climatic episodes are generally recognized:

1   the Medieval Warm Period (AD 900–1300; Lamb, 1982);
2   the Mid-millennium Climatic Deterioration, otherwise known as 'The Little Ice Age' (AD 1300–1900); and
3   the twentieth-century Warming, or Global Warming, Epoch (AD 1900 onwards) due to the human-accentuated greenhouse effect.

Thus the millennium starts and finishes with warm episodes, with temperatures in the final two decades conforming to those experienced during the Medieval Warm Period, when human occupancy and farming pushed progressively uphill to reach elevations of over 400 m in western and northern Britain, as testified by evidence from Dartmoor and the Lammermuir Hills (Lamb, 1985). However, the majority of the last millennium was characterized by temperatures well below those of the present. For 600 years, temperatures were generally depressed in Britain and Europe more generally.

Significantly less is known about rainfall patterns because of the profound affects of local and regional factors. However, both total rainfall and winter rainfall appear to have been lower in the colder periods. The best record of surface wetness remains the curve produced for Bolton Fell Moss (Barber, 1981), which shows prevailing wet to very wet conditions except for the three warm phases; prior to AD 1300, AD 1470–1560 and after AD 1900 (the fourth brief and relatively insignificant 'less cold' phase from AD 1700 to AD 1740 is not identified). However, whether this wetness is due to increased precipitation, changes to the seasonality of precipitation or reduction in evapotranspiration at this site remains unclear. Archives also provide evidence of windiness in terms of storms (Lamb, 1985). These reveal that there were phases of increased storminess during the transitional stages from the Medieval Warm Period to the Little Ice Age (AD 1200–50 and AD 1300–50), as well as during the coldest periods of the millennium (1400–60, 1560–1610, 1670–1710 and 1790–1810). Conditions did vary significantly within the period 1300–1900, but it must be recognized that the evidence is limited prior to 1700. It also has to be recognized that changes were not necessarily synchronous across Britain, and that much has yet to be discovered about the true magnitudes and patterns of climatic change during this period.

3.3   Temporal Influences on Natural Slope Processes: Mass Movement

Historical records studied by Grove (1988) and others have clearly revealed that the climatic deterioration associated with the onset of the Little Ice Age in the Alps and Scandinavia undoubtedly led to an increased frequency of avalanches, rockfalls, landslides and bog bursts. In particular, the study of parish records in Norway (Grove, 1972, 1988) emphasizes a phase of great activity dating from AD 1685 to 1765. Such evidence clearly suggests that the same may have happened in Britain; at least in Scotland, which is thought to have had a recent climatic history relatively similar to that experienced in Norway because of the more direct exposure to the same warming and cooling episodes of the northern Atlantic (Lamb, 1985).

In reality, the evidence is far from conclusive. Price (1983, 186–7) writing about landform evolution in Scotland, presented a traditional view by stating that, 'the minor changes in temperatures (plus or minus 1°C) and precipitation (plus or minus ten per cent) have had no significant impact on vegetation, soils and landforms over the last 4000 [years]', although he did go on to add, 'This is not to say that a succession of cold winters (e.g. in the 17th century AD) did not increase the amount of periglacial activity above 800 m . . .'. This view appeared to be supported by the much-quoted study by Innes (1983) into debris flow activity in the Scottish Highlands, which showed that debris flows were predominantly a feature of the last 250 years, stimulated by the human use of fire in vegetation management. However, there are now grounds for questioning the objectivity of these results with regard to the frequency and extent of debris flow activity prior to AD 1750, and so the results are of no use in clarifying the role of the Little Ice Age in landslide generation.

A similar situation exists with reference to southern Britain, where the deterioration in climate is thought to have been rather less severe. A review by Hutchinson (1965) reached the conclusion that there had indeed been increased landslide activity during the period 1550–1850, most especially from 1550 to 1600 and from 1700 to 1850. The former coincides with one of the severest episodes of the Little Ice Age, and is known especially for the famous 'Wonder Landslide' at Marcle, Herefordshire, which occurred on 17 February 1575 and took the form of a massive translational block-glide (60 000 m³), produced through a combination of denudational unloading and high pore-water pressures (Hall and Griffiths, 1999). As the latter episode is both longer and more recent, it comes as no surprise that there is a better record of conspicuous failures. Included are the most serious snow avalanche recorded for England, which descended on a row of houses at Lewes, Sussex, on Christmas Eve, 1836, killing eight people; a number of bog bursts, most noticeably that affecting Solway Moss in 1772 (Lyell, 1835); and several major landslide events, including the failure at Whitestone Cliff in the Hambleton Hills (1755), the Birches slide in the Ironbridge Gorge (1773), the Hawkley slide, near Selborne (1774), the Beacon Hill landslide at Bath (1790) and the failure of the north-east slopes of the Isle of Portland (1792). By contrast, a review of major slope failures undertaken by Johnson (1987) concluded that major rockslides have not occurred with any marked periodicity, and that the high rainfall events which trigger such mass movements need not be associated with particular climatic periods.

It was hoped that the desk study survey of landsliding in Great Britain, commissioned in the mid-1980s by the then Department of the Environment and completed in 1991, would further clarify the situation. Using all extant published information accessible within the public domain, the survey revealed a total of 8835 reported landslides in Britain, 7533 of which were 'non-coastal' (table 3.1). It must be

Table 3.1 The relative frequency of landslides of different ages in Great Britain, as revealed by the Department of the Environment sponsored review of research into landsliding in Great Britain (after Jones and Lee, 1994)

	Inland		*Coastal*		*Total*	
	No.	*%*	*No.*	*%*	*No.*	*%*
Active	296	3.9	483	37.1	779	8.8
Recent	927	12.3	423	32.5	1350	15.3
Relict	970	12.9	66	5.1	1036	11.7
Fossil	669	8.9	45	3.5	714	8.1
Unknown age	4671	62.0	285	21.8	4956	56.1
TOTAL	7533		1302		8835	

noted that the distribution map (figure 3.3(a)) merely shows the location of known (i.e. 'reported') failures and not the actual distribution of extant landslides on the ground. It is, therefore, a reasonable representation of current knowledge of landslide distribution and, as such, reflects how and where landslides have been investigated in the past (see Jones and Lee (1994) and Jones (1999) for full discussions of the limitations of the data).

In order that the survey should yield an appreciation of the pattern of landsliding through time, a five-fold age classification scheme was adopted at the outset:

Active: sites that are currently unstable, cyclically unstable with a periodicity of up to five years, or first-time and reactivated failures that occurred within five years of the commencement of the survey (i.e. 1980–1987).

Recent: slope instability sites known to have suffered movements within the last 100 years.

Relict: sites known to have suffered movements within the historical timescale (i.e. 100–1000 years) where landforms are still clearly recognizable but show no obvious signs of contemporary movement.

Fossil: indicating instability features that developed in early historical and prehistoric times.

Unknown: indicating that no statement on age was provided in the source material examined.

In reality, the nature of the published record of landsliding often made it difficult to distinguish between 'Active' and 'Recent' failures as

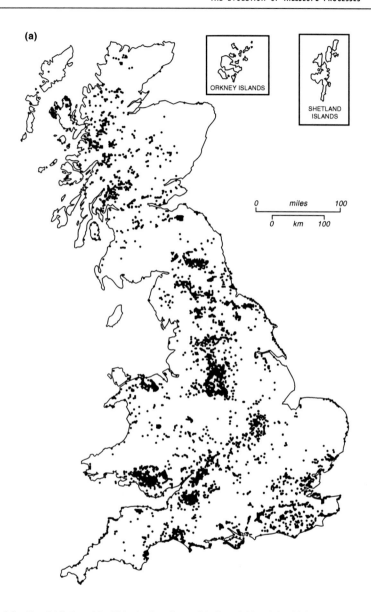

Figure 3.3 The distribution of landslides in Great Britain (a). 'Active' (b) and 'youthful' (c) landslides in Great Britain (after Jones and Lee, 1994).

well as between 'Relict' and 'Fossil' landslides, resulting in the gradual replacement of the original four-fold division with two new broader groupings, 'Youthful' and 'Ancient' landslides. Nevertheless, the data

(b)

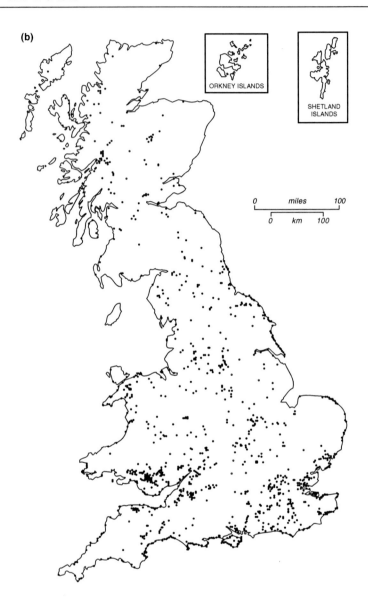

ORKNEY ISLANDS

SHETLAND
ISLANDS

Figure 3.3   *continued*

collection exercise revealed that a total of 2193 inland landslides are
recorded as having occurred during the last millennium, representing
29.1% of the non-coastal data set, of which a total of 970 are con-

(c)

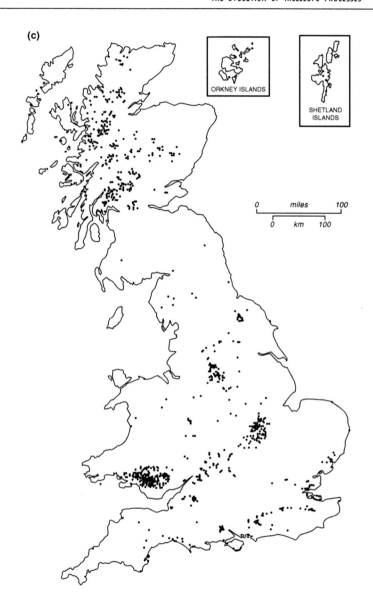

Figure 3.3 *continued*

sidered to have occurred more than a century ago (figure 3.3(b),(c)). It should also be noted that only 669 inland failures are reported to be over 1000 years old and that a staggering 4671 inland landslides, representing a dominant 62% of all inland landslides (i.e. landslides identified in the survey) recorded, are of unknown age, indicating the clear need for further research on past landslide activity in Britain. As a significant proportion of the latter almost certainly moved during the last 1000 years and 'known' landslides only represent an uncertain proportion of all landslides that have occurred in this period, it can safely be assumed that there have been a great many more than the reported 2193 landslides in the last 1000 years.

While the survey presents ample evidence for slope failures during the last millennium, it also reveals that the dating of many failures remains vague, and so the role of the climatic deterioration during the Little Ice Age continues to be unclear. It is interesting to note that many of the geomorphologists who undertook the primary regionally based literature searches to provide the basic inputs into the landslide survey made little or no mention of the Little Ice Age in their reports, which were submitted in the mid-1980s. Of rather greater importance is the fact that the famous translational slides on the Lower Greensand escarpment to the south of Sevenoaks, reactivated during the construction of the A21 Sevenoaks Bypass in 1966 and subsequently the subject of detailed investigation, show little sign of movement over the last 10 000 years (Skempton and Weeks, 1976). Similarly, the more recent work by Hutchinson et al. (1985) on the Roman Fort at Lympne in Kent, whose much disturbed remains are located on the landslide-prone abandoned cliffline that forms the northern margin of Romney Marsh, has revealed that following its abandonment in AD 340–350, it was severely disrupted by landsliding in 540–590, with subsequent, shallower movements at some stage after 700. Although there is a record from 1724 of a local rotational failure on the slope crest about half a kilometre to the west, no evidence has yet been found to link these later, shallower failures with the Little Ice Age phase of climatic deterioration.

However, the records from the time do indicate cold and damp conditions, with lengthy accumulations of snow. For example, Gilbert White (1789) wrote of the Hawkley failure, 'The months of January and February, in the year 1774, were remarkable for great melting snows and vast gluts of rain. The beginning of March also went in the same tenor; when on the night between the 8th and the 9th of that month, a considerable part of the great woody hanger at Hawkley was torn from its place.' Similarly, Hutchins (1803), writing of the 1792 Isle of Portland failure, stated 'the season had been very wet . . . the week preceding these there had been some strong gales of wind.' It is quite possible, therefore, that the cooler, wetter conditions that appear to have characterized much of the millennium could have elevated

groundwater tables and increased pore-water pressures, thereby contributing to increased landslide activity, including the generation of a
number of relatively large, conspicuous failures. The main problem is
that individual and sequences of weather events capable of triggering
slope instability are frequently localized, as is the distribution of slopes
susceptible to slope failure (i.e. displaying marginal stability) at any
particular point in time, so their coincidence is often a matter of
chance. For example, according to the England and Wales Precipitation
record, both of the two failures mentioned above occurred in quite
average years. Thus it is quite possible for evidence of the role of the
mid-millennium climatic deterioration in generating slope instability to
be present at some sites and on some rocks and surficial deposits, and
yet be absent elsewhere. This would go some way towards explaining
why the current evidence for 'Little Ice Age' failures is best preserved
on the Jurassic outcrops of the Midlands, where Chandler (1970)
identified periodic mudsliding and slips post AD 1660 in the Gwash
Valley, Rutland, and Butler (1983), in a wide-ranging survey of landsliding in the Cotswolds, concluded that there had been major localized rotational failures and mudsliding along the escarpment over the
period 1500–1820. In the case of the latter, particularly dramatic
evidence is to be found on the flanks of Ebrington Hill, Warwickshire,
where mudslide lobes can be seen to override medieval ridge-and-
furrow agricultural patterning and to disrupt enclosure field boundaries
established in 1810. However, as both the ridge-and-furrow and the
1810 enclosures also pass across pre-existing lobes, the question
remains as to just how significant the mid-millennium climatic deterioration really was in terms of the generation of slope instability, for the
features identified could represent the final stages of postglacial slope
evolution or the product of extreme rainfall events which occur irrespective of the prevailing climatic regime (see Johnson, 1987).

Such uncertainties are further highlighted by two surveys of extreme
events published in recent years. The first is Lamb's (1991) analysis of
storm severity over the period 1570–1989, while the second is a partial
survey of the frequency of erosional, depositional and flooding events
from 1700, prepared as part of a report entitled *Erosion, Deposition and
Flooding* (Lee, 1995). Both reveal the characteristic increase in record-
ings to the present, which is mainly due to a combination of progressively increasing human awareness, improved record-keeping and better
record survival. The record of storminess (Lamb, 1991) reveals storm
occurrences throughout the period, but three episodes of particularly
severe activity: prior to AD 1650, corresponding to the first cold phase
of the Little Ice Age, when surface wind strengths are thought to have
exceeded those of recent times, and are notable for the Great Breckland

Duststorms which occurred over the years 1570–1688, 1880–1900 and, finally, from 1950 onwards. The survey contained in Lee (1995), while even less revealing about changes in climate and climatically induced events, does highlight the important and ever-growing interaction between human activity and the physical environment, and the need to consider both in combination in any general review of slope processes over the past millennium. Thus the potential effects of the Little Ice Age were undoubtedly greatly influenced by the nature and extent of human modifications to landscape. In some cases human activity and climate would have combined to exacerbate slope instability, while in other instances the reverse may have been true, and it is quite possible that some of the products of these times have been obscured, or even erased, by the impacts of human activity over the last two centuries. It is clear, therefore, that it is first necessary to examine the character of human impacts before returning to the topic of slope evolution over the millennium.

3.4 The Human Impact on Landforms

The role of humans in creating geomorphological change has received increasing attention in recent years and general reviews are to be found in Nir (1983) and Goudie (1995, 2000), including discussions of origins and morphological characteristics. A general introduction to the situation in Britain is outlined in Goudie (1990) but despite the existence of several illuminating discussions on the evolution of the British landscape (Hoskins, 1955; Woodell, 1985; Rackham, 1986), there is as yet no coherent consideration of the human impact on surface form through time and over space. It is widely recognized that human activity can affect surface form either directly or indirectly, although classification schemes and terminologies vary. This chapter adopts the fundamental division into *human-made, human-modified* and *human-induced* categories, as will be explained. There is, generally, little disagreement with regard to the broad category of *human-made* or *human-created* landforms, which are, typically, excavations and mounds of enormous variety created directly by human action. However, a rather more contentious three-fold subdivision can be recognized – 'purposeful', 'incidental' and 'accidental' – to reflect the extent to which the shape of the resulting landform had been predetermined or was merely a by-product of activity. Human-made landforms are predominantly purposeful and achieve a huge variety of forms, including quarries, pits, burial mounds, road and railway embankments, spoil heaps and tracts of reclaimed land. Sometimes referred to as 'artificial' landforms, these features lie at the very heart of anthropogeomorphology, for they are created like natural landforms through the processes of erosion, transportion and deposition although, unusually in this instance, the nature of the geomorphological agents has changed through time as a consequence of the harnessing of various kinds of energy and the development of technology. Thus human-made landforms have displayed marked evolution through the millennium in terms of their form, magnitude, frequency, extent and rapidity of creation, most especially since the Industrial Revolution. Indeed, the increasing tendency in recent decades to create 'engineered landscapes' has led some to refer to the process of 'bulldozergenesis'.

Purposeful human-made landforms are now ubiquitous in Britain. Conspicuous examples dating from the early pre-Industrial-Revolution part of the last millennium include the Norfolk Broads, a collection of 25 freshwater lakes occupying excavations created by the removal of 25.5 million cubic metres of material in peat diggings prior to AD 1300, and the undulating ridge-and-furrow surfaces that used to cover much of the English Midlands, produced by medieval ploughing practices and involving the movement of an estimated 62 000 m³ of material per square kilometre. There are also the ubiquitous small pits produced throughout the period for a multitude of reasons, many now filled with waste or spoil to form 'made ground', but others still surviving as small copses and ponds. No fewer than 37 000 pits were recorded in Norfolk in the 1950s (Prince, 1964), while Rackham (1986) reports the existence of an estimated 388 000 small ponds in England and Wales, based on a survey of maps dating from the 1920s, and suggests that the total may have been 800 000 in the 1880s. However, these features pale into insignificance when compared with the scale of earth movement and surface reconfigurement undertaken in the last two centuries of the millennium. For example, the expansion of the railway network in the mid- and late-nineteenth century led to prodigious anthropogeomorphic activity in terms of embankments, cuttings and tunnels; practices continued at the present in the ever-expanding road and motorway network. Similarly, the expansion of coal-mining led to dramatic increases in spoil production which, through the use of mechanical tipping, was deposited in hundreds of conical spoil heaps standing up to 50 m high, thereby creating the characteristic coalfield landform of the later millennium (e.g. the 'Wigan Alps'). Indeed, the millennium witnessed the creation of hitherto unknown landforms in terms of linear trough–ridge systems, the tunnel and the accumulation cone, and there are undoubtedly some who would wish to add buildings and bridges made of earth materials to the list, especially since the development of the 'green bridge' concept.

Recent developments in ground engineering earthworks design and in landscape architecture have resulted in greater and greater attention being directed to creating specific landform morphologies, thereby raising the possibility of distinguishing between 'purposeful' and 'incidental' human-made landforms. The fact that humans did not explicitly set out to create conical spoil heaps and other 'dumped' depositional forms but that they arose by chance from some other endeavour (e.g. mining for coal) suggests that they may be better termed *incidental* human-made landforms. In this regard, the conical spoil heaps were certainly different to the growing number of low convexo-concave hills recently produced through the disposal of domestic/commercial waste (Gray, 1997, 1998), which are purposefully fashioned in this form. It is possible to argue that it was only when conical spoil heaps were refashioned from the 1970s to the 1990s in order to blend better with surrounding landscapes, or to create amenity sites, that they truly became *purposeful* human-made landforms. Perhaps the best example of 'incidental' human-made landforms is the general accumulations of detritus within urban areas, which Sherlock (1931) estimated to be between 3 and 8 m thick beneath the City of London.

Within the spectrum of human-made landforms, the rather debatable 'incidental' category lies intermediate between the 'purposeful' and 'accidental' groups. *Accidental* human-made landforms are created unintentionally and include craters, produced by bombs, accidents and explosions. This three-fold division of human-made landforms clearly has deficiencies, and it may prove more satisfactory in the future to distinguish between 'designed' and 'non-designed' forms. Human activity also affects environmental conditions and thereby the operation of geomorphological processes and landform development. These indirect affects can result in the creation of *human-induced* and *human-modified* landforms. *Human-induced* landforms are created by natural processes, but in locations and at times that are wholly dependent on human activity. Such landforms include gullies developed on spoil heaps, the deltas and beaches associated with impounded reservoirs and the subsidence features arising from the underground extraction of coal, iron ore and salt. It is the last mentioned that is most widespread in Britain, mainly associated with past coal-mining, where it can locally exceed 10 m (Sherlock, 1931), but most dramatically in Cheshire, where solution mining of two shallow 30 m thick seams of salt has resulted in the creation of numerous clearly defined depressions, many of which contain lakes known as 'flashes' (Wallwork, 1974).

The remaining category of *human-modified* landforms is created when the distribution and/or rate of operation of geomorphological processes are changed as a consequence of human activity. The classic cause is a changed surface hydrological budget, arising from alterations in the nature and distribution of vegetation cover, agricultural practices and construction activity, although climate change due to human-induced global warming may well qualify in the future. Numerous examples exist of dramatically raised sediment yields from such activities. Sometimes the changes in land use can result in rapid and profound landform modification – as, for example, when rapid deforestation results in the accelerated removal of soil, thereby leading to rill and gully development and the creation of shallow translational slides. The resulting landforms are easy to delimit and measure, as is also the case with gullies developed along pathways in heavily used recreational areas or in the network of holloways (sunken lanes) created in the ancient countryside of lowland Britain. However, for the most part the modifications to landforms are difficult to detect, except where obstructions, such as hedgelines, reveal differences in elevation due to accelerated creep. Thus, slowly accumulated changes often go undetected and in a great many instances it is impossible to identify, let alone quantify, how much morphological change has actually resulted from human influences.

Detailed investigations, including monitoring, have produced mounting evidence for the relatively recent acceleration of soil erosion in Britain, both by wind and water, arising from changes in agricultural practices such as changing crops, mechanization and the removal of hedges and copses to facilitate expansive monoculture. One result has been the increased occurrence of 'soil blows' and duststorms in the windier late twentieth century. Duststorms are not new to Britain and have been recorded in historical times from most areas with light sandy soils, most especially in the 'Great Duststorms' of the Brecklands during the period 1570–1688, when 'sandbanks near 20 yards high' were produced (Wright, 1668) and the town of Santon Downham was gradually engulfed by moving sand in 1630, including blocking the river for 5 km (see Lee, 1995). The area has since been stabilized by afforestation, but the events highlight the scale of destruction that can occur in the severest of storms, if sandy soils are left exposed. Other severe events were reported from Bagshot Heath by Daniel Defoe (1724–6) and from Norfolk (Reid, 1884). Another memorable series of events was the Culbin Sands disaster of 1694 and following years. At that time, 16 fertile farms covering some 20–30 km² near Findhorn and Forres on the Moray Firth were overwhelmed in a single violent storm. The whole area, including the mansion house, was buried by up to 30 m of loose sand. From 1694 to 1704 there were frequent periods of blowing sand, and the area remained a desert of shifting sand for 230 years until the Forestry Commission successfully afforested it during the 1920s. Much of the damage is believed to have been the result of a single violent storm, probably in late September or October 1694:

> At first only fields were invaded (by the sand). A ploughman had to leave his plough, while reapers left their stooks of barley. When they returned, both plough and barley were buried for ever. The drift then advanced upon the village, engulfing cottages and the laird's mansion. The storm continued through the night, and next morning some of the cottars had to break through the backs of their houses to get out. On the second day of the storm, the people freed their cattle and fled with their belongings to safer ground. Their flight (southeastwards or eastwards) was obstructed by the river Findhorn: since its mouth had been blocked by the drifting sand, its waters rose until it could force a new passage to the sea. (original account quoted from Edlin, 1976).

The village of Culbin was buried by the storm, although over the following centuries buildings have been released from the sand only to disappear again. It is believed that the erosion on the Moray plain was the result of winds of 100–130 knots (50–65 m s^{-1}) blowing inland from the NNW, accompanied by spring tides which probably caused the sea to erode the dune faces, exposing the dry sand cores to wind scour. Another factor may have been the destabilization of the dunes by harvesting of marram grass for local industries. Wind erosion damage has once again become increasingly recorded since the 1920s from a wide range of localities, including the Fenlands, Brecklands, East Yorkshire, Lincolnshire and the Midlands (Robinson, 1969; Wilson and Cooke, 1980; Fullen, 1985). As a consequence, over 6000 km^2 of England and Wales is now considered prone to wind erosion (Morgan, 1985; Evans, 1990), with recorded losses of up to 6–10 t ha^{-1} in particularly bad years (Fullen, 1985).

The increase in soil loss due to surface run-off from arable fields is even more startling, although rates remain generally low. Changes in land management and crop production which leave the soil exposed in the winter months have resulted in reports of greatly increased erosion. Although dramatic features such as gullies, crop burials and mudfloods are relatively limited in their occurrence (Boardman, 1990a, 1992), there have been reports of housing being impacted (Boardman 1990b) and it is now estimated that 37–45% of arable land in England and Wales is susceptible to erosion (Morgan, 1985; Evans and Cook, 1986), with anything from 5% to 15% suffering appreciable loss in any one year (Evans, 1988). Typical annual rates of soil loss are estimated at 1–5 t ha^{-1}, but estimates of maximum losses in extreme events have risen from 17–18 t ha^{-1} in the 1970s to between 50 and over 100 t ha^{-1} (Evans, 1988; Boardman, 1990a).

Land drainage to improve agricultural productivity and alleviate flooding problems has led to marked subsidence in areas of peat soils (peat wastage). As the water table was lowered, so the peat became susceptible to oxidation and deflation. In the Fens, for example, significant surface lowering quickly followed the land improvements started by Vermuyden in the seventeenth century. Concerns about the resulting engineering and economic problems led a landowner, William Wells, and his engineer, John Laurence, to establish a firm datum in Whittlesey Mere (south of Peterborough) prior to its drainage in 1850. This datum, the Holme Post, has indicated that around 4 m of surface lowering has occurred since that date. The average wastage rate in the first 10 years after drainage was 177 mm yr^{-1}.

By way of commencing a conclusion to this section on human influences on slope processes and forms, it is worth restating that these influences have become increasingly diverse, complex and dominant as the millennium progressed. Unfortunately, it remains impossible fully to appreciate the magnitude of these influences, as the basic quantitative research remains to be done (see Lee, 1995). Only in those instances where pre-existing landforms have been significantly and obviously altered, new 'atypical' landforms developed or distinctive 'artificial' landforms created is it possible to gain some appreciation of the scale of change. But, even in these cases, great care needs to be exercised. While some landforms tend to retain their identity due to their shape, position or composition, and only respond slowly to change (what may be termed 'robust' landforms, following Werrity and Brazier, 1994), others may be much more susceptible to change (akin to the 'responsive' landforms of Werrity, 1997) and easily become obliterated. Thus the surface of Britain may be envisaged as covered with landforms or differing size, composition and origin, which display greatly varying degrees of modification during the millennium due to natural, quasi-natural (human-modified and -induced) and direct human action. As estimated mean denudation rates for Britain based mainly on river sediment yields reveal values of less than 40 mm per 1000 years, it is clear that some hard rock landforms remote from human influences may well have suffered imperceptible lowering through weathering, while adjacent examples may have been significantly modified by rockfall activity and so on. Similarly, overall solution rates suggest that limestone terrains may have suffered considerably less than 50 mm solutional lowering over the same period, although local values may be much higher. By contrast, current estimations of typical soil erosion by water (see above) equate to mean surface lowering rates of 38–190 mm per 1000 years, suggesting some quite high values in particularly vulnerable areas. However, the really large values for natural change over the millennium would be in areas subjected to mass movement activity, where local changes in elevation of up to 50 m would result from rotational failures, rockslides and rockfalls, and the resulting accumulation of debris.

But how do these changes compare to those created by direct human activity? Certainly, the largest changes in elevation are due to humans, in terms of quarries, spoil heaps, cuttings and embankments. In terms of overall geomorphological activity, there are, as yet, no precise figures for earth movement due to humans. Sherlock (1931) estimated that by 1913 humans had excavated $c.31$ billion m^3 of material in Great Britain, mainly from mines (50%), pits and quarries (39%) and associated with the construction of the railways (7.6%), and he went on to state, 'Man is many more times more powerful, as an agent of denudation, than all the atmospheric denuding forces combined' (Sherlock, 1931, 333). Anthropogeomorphic activity has undoubtedly increased since 1913, for while production of deep-mined coal has fallen dramatically, there has been a huge rise in the surface extraction of industrial minerals/ construction materials (sand and gravel, limestone, clay and sandstone), so that surface pitting and quarrying has grown from $c.40$ million tonnes in 1900, to 115 Mt in 1938, 192 Mt in 1955, 300 Mt in 1967 and 306 Mt in 1986 (Jones, 1991). As these figures exclude spoil production, it would appear to be safe to conclude that the surface mineral extraction industry was moving at least 400 Mt of earth materials per year in the mid-1980s, and that the total earth-moving activity was probably double that value at 800 Mt yr^{-1}. Using an average density of 2.2, this works out at roughly 365 million m^3 of material, or 1585 m^3 km^{-2}, for the land area of Great Britain, equivalent to a surface lowering of 15.85 mm per annum. Clearly, all this activity results in both surface lowering and raising. It also cannot be simply compared with sediment yields, which are measures of output and not of work done. Nevertheless, the net calculated contemporary fluvial denudation rates, which usually fall in the range 30–100 t km^{-2} yr^{-1}, appear insignificant when compared with the estimated human activity rates of 3485 t km^{-2} yr^{-1}, or even surface mineral extraction in 1986 (1333 t km^{-2}), thereby confirming Sherlock's (1931) conclusion.

Returning consideration to the surface form of Great Britain, it is possible to envisage it as composed of a mosaic of landforms, each one of which can be placed on the spectrum 'completely natural' to 'completely artificial', using an adaptation of the scheme proposed by Graf (1996) for rivers in the United States (table 3.2). At the start of the Holocene, all landforms in Britain would have been classified as

Table 3.2 A proposed classification scheme of geomorphic 'naturalness' (based on Graf, 1996)

	Completely natural	Essentially natural	Partly modified	Substantially modified	Mostly modified	Essentially artificial	Completely artificial
Major landforms	No obvious evidence of human activities – same form and process as existed prior to human occupation	No obvious evidence of human activity – same forms as prior to human occupation	Locally evident changes as a result of human activities	Significant changes due to human activity, in terms of human resculpturing	Widespread alteration as a result of human activities; much resculpturing and in some areas difficult to identify pre-human morphology	Predominantly altered as a result of human activities; often difficult to establish pre-human modification form	Completely engineered
Minor landforms	Same forms and processes as those found prior to human occupation	Some few signs of alteration by human activities or changes in erosion and sediment supply	Some human-made landforms and alterations to landforms; some changes due to alterations in erosion or sediment supply	Many altered by human activity and many more replaced by human-made forms; good evidence of human-modified landforms	Few wholly natural forms survive intact	Very few natural forms survive intact	All artificial
Percentage of area engineered, disturbed or detectably altered through human activity	0%	<10%	<20%	>20% <50%	>50% <90%	>90% <100%	100%

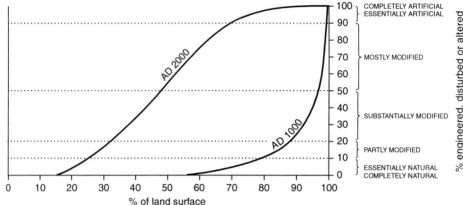

Figure 3.4 A tentative portrayal of the changing degree of naturalness of the landforms of Great Britain.

'completely natural', but as human societies evolved and impacts on surface form became more and more significant and extensive, so the proportion of land attributed to the various 'modified' and 'artificial' categories has grown at the expense of the 'natural' category. This undoubtedly happened in Britain over the last millennium (figure 3.4), as is discussed in the next section. But, at the same time, there have been interesting developments in terms of the evolution of human influences on landform development (figure 3.5). First, there is clear evidence of progressively increased intensity of human impacts, resulting in the creation of artificial landforms of greater and greater size and extent. Second, there is evidence of cycles of landform sculpturing, so that forms created at one time are subsequently refashioned for new purposes (e.g. colliery spoil heaps). Third, there is clear evidence for convergence, in that human-created forms frequently mimic natural forms – a particular feature of the last three decades of the twentieth century due to the rise of landscape architecture, fuelled by the desire to recreate the natural or, at the very least, to produce forms that appear natural to the general public. Finally, modified and artificial forms are re-sculptured by 'natural' processes so as to attempt to complete the cycle, although it is unusual for natural forms to result because of further human activity.

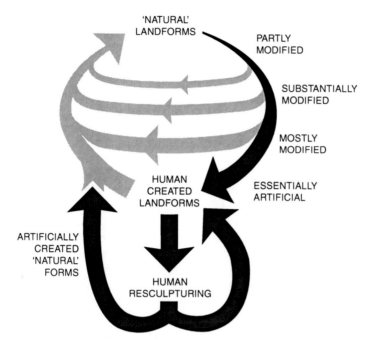

Figure 3.5 A simplified portrayal of anthropogeomorphic cycles. It should be noted that the modification of existing landforms by natural processes is capable of being arrested/reversed by further human influences (either direct or indirect), thereby indicating the existence of further return pathways not shown on the diagram.

3.5   Slope Processes and Evolution Over the Millennium

Drawing on the above discussion, it is possible to put forward a tentative outline of the major influences on slopes and slope processes over the millennium.

It is now widely accepted that humans had already considerably altered the landscape of Great Britain by AD 1000, largely through fairly extensive forest clearance in lowland areas. Such clearance activity, together with the adoption of primitive agricultural practices, would have undoubtedly resulted in the patchy occurrence of accelerated soil erosion and limited slope instability for up to 5000 years previously. The actual timing of events and the magnitudes of impacts would have varied greatly from area to area, with some recording no impact whatsoever. As a consequence, no generalization has yet emerged, although the overall fluctuations in erosion may well have corresponded to the model for southern Sweden put forward by Dearing (1991).

After AD 1000, overall erosion may be envisaged to have increased during the Medieval Warm Period as forest clearance progressed and agriculture expanded, and the effects may have been particularly profound in those upland areas which were being properly settled for the first time.

What happened during the 'Little Ice Age', broadly defined as AD 1300–1900, is more complex and remains difficult to determine. The coldest periods could have stimulated renewed free–thaw activity in highland areas. In upland regions, the retreat of humans obviously reduced the scale of human modification of landforms by limiting accelerated soil erosion through surface run-off, but these influences may have been offset by the cooler, damper conditions and increased snowfall and snowmelt. In lowland areas, it appears likely that erosion continued to increase because of a combination of population growth and the deterioration in climate, and there is evidence of increased slope instability in some areas. Overall, therefore, erosion is likely to have increased steadily until the eighteenth century.

The influence of climatic deterioration on geomorphological activity becomes increasingly obscured from the eighteenth century onwards, because of human activity. The onset of the Industrial Revolution was gradual, but its effects on landscape were to be profound. Hoskins (1955, 211) noted that 'England was still a peaceful agricultural country at the beginning of the seventeenth century' and that the effects of the first industrial revolution had, as yet, left 'little to show for it in the landscape' and 'so far as visible signs upon the face of the country were concerned it was all a mere scratching of the surface'. However, Hoskins goes on to state that 'By the end of the seventeenth century the industrial landscape was much more evident' (Hoskins, 1955, 211) and continued technological development, population growth and population redistribution were to progress the transformation of the landscape at an ever increasing pace. The ability to achieve geomorphological work must also be considered to have increased several-fold per head of population during the millennium. Thus England steadily moved from 'being a country of tightly-packed towns in 1800 to being in 1900 the most profligate country in the industrialised world in its use of land for housing' (Thompson, 1985, 173). Ground disturbance became increasingly widespread and intense, with the result that erosion rates rose dramatically from the eighteenth century onwards.

But the nature of the human impact also changed significantly over the period AD 1000–1900. Prior to the onset of the millennium, the human impact had been restricted to numerous human-made landforms of relatively limited extent (burial mounds, linear earthworks,

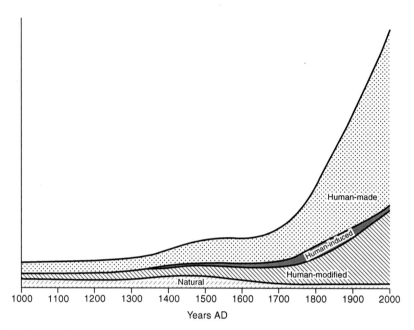

Figure 3.6 A preliminary model of the human impact on landforms over the last millennium.

hill-forts, settlement sites and so on), although there were well-developed concentrations (e.g. Salisbury Plain), and to more extensive areas of human-modified forms that displayed relatively limited degrees of alteration. Thus the overwhelming majority of the land surface of Britain could still be described as 'completely natural' or 'essentially natural' (table 3.2). During the period between 1000 and c.1750, the creation of 'human-made' landforms increased dramatically, as did their scale and extent. This, in turn, led to the development of limited numbers of 'human-induced' landforms, while the ever-growing transformation of landscape led to the increasing extent and significance of 'human-modified' landforms. Thus the three populations grew at variable rates (figure 3.6) and at the expense of natural forms. During the period 1750 to 1900, this process accelerated dramatically, but most especially in terms of the creation of 'human-made' forms, which became ubiquitous in many industrial areas, resulting in the creation of tracts of wholly re-sculptured 'artificial' land surfaces.

During the last century of the millennium, otherwise known as the Global Warming Epoch, these trends have continued. Anthropogeomorphic activity remains as high as it ever was and extensive tracts of land are re-sculptured by humans each year, sometimes in an attempt to recreate natural forms. Human-induced and human-modified forms also continue to evolve at pace, with the inevitable consequence that the extent of 'completely natural' and 'essentially natural' forms continues to decrease, and to be more and more restricted to the outcrops of resistant rocks in the north and west.

To conclude, therefore, a review of influences on slope processes and slope evolution over the millennium reveals the following:

- There is a long history of human influences on landform development in Great Britain that considerably predates AD 1000. By AD 1000 humans had already had a profound impact on landscape, especially in lowland areas, and some effect of landforms.
- Human influences on slope form and process have increased over the millennium and, from the perspective of Great Britain as a whole, have attained a dominant role since the mid-eighteenth century, thus raising the possibility of calling post-1750 the *Anthropocene*. However, some more remote areas continue to evolve under an essentially natural regime.
- Direct human influences on landform development have diversified dramatically over the millennium, resulting in increases in the variety, scale and extent of resulting landforms. It is possible to identify an evolution of human-made landforms involving the introduction of completely 'new' landforms, cycles of landform development, convergence and pseudo-natural or mimicked landforms.
- Much work remains to be done on identifying the significance of human-modified landforms in Britain and in quantifying the role of humans in landform creation.
- The significance of the Little Ice Age, in terms of slope evolution in Britain, remains to be established, as does the role of extreme events.

REFERENCES

Ager, D. V. 1973: *The Nature of the Stratigraphic Record*. London: Macmillan.
Barber, K. E. 1981: *Peat Stratigraphy and Climate Change*. Rotterdam: Balkema.
Boardman, J. 1990a: *Soil Erosion in Britain: Costs, Attitudes and Policies*. Social Audit Paper No. 1, Education Network for Environment and Development, University of Sussex.
Boardman, J. 1990b: Soil erosion on the South Downs: a review. In J. Boardman, I. D. L. Foster and J. A. Dearing (eds), *Soil Erosion on Agricultural Land*. Chichester: Wiley, 87–105.

Boardman, J. 1992: Agriculture and erosion in Britain. *Geography Review*, 6, 15–19.

Brown, E. H. 1979: The shape of Britain. *Transactions of the Institute of British Geographers*, NS, 4, 449–62.

Burton, I. and Kates, R. W. 1986: The Great Climateric, 1748–2048: the transition to a just and sustainable human environment. In R. W. Kates and I. Burton (eds), *Themes from the Work of Gilbert F. White*. Chicago: University of Chicago Press, 339–60.

Butler, P. B. 1983: Landsliding and other large scale mass movements on the escarpment of the Cotswold Hills. Unpublished B.A. dissertation, University of Oxford.

Chandler, R. J. 1970: The degradation of Lias Clay slopes in an area of the East Midlands. *Quarterly Journal of Engineering Geology*, 2, 161–81.

Dearing, J. A. 1991: Erosion and land use. In B. E. Berglund (ed.), *The Cultural Landscape during 6000 years in Southern Sweden. Ecological Bulletins*, 41, 283–92.

Edlin, H. L. 1976: The Culbin Sands. In J. Lenihan and W. W. Fletcher (eds), *Environment and Man*, Vol. 4. *Reclamation*. London: Blackie, 1–13.

Evans, R. 1988: *Water Erosion in England and Wales, 1982–1984*. Report for Soil Survey and Land Research Centre, Silsoe.

Evans, R. 1990: Soils at risk of accelerated erosion in England and Wales. *Soil Use and Management*, 6, 125–31.

Evans, R. and Cook, S. 1986: Soil erosion in Britain. *SEESOIL*, 3, 28–58.

Fullen, M. R. 1985: Wind erosion of arable soils in East Shropshire (England) during spring 1983. *Catena*, 12, 111–20.

Graf, W. L. 1996: Geomorphology and policy for restoration of impounded American rivers: What is natural? In B. L. Rhoads and C. E. Thorn (eds), *The Scientific Nature of Geomorphology: Proceedings of the 27th Binghamton Symposium in Geomorphology*. Chichester: Wiley, 443–73.

Gray, J. M. 1997: Planning and landform: geomorphological authenticity or incongruity in the countryside. *Area*, 29, 312–24.

Gray, J. M. 1998: Hills of waste: a policy conflict in environmental geology. In M. R. Bennett and P. Doyle (eds), *Environmental Geology: a British Perspective*. London: Geological Society, 173–95.

Grove, J. M. 1972: The incidence of landslides, avalanches and floods in Western Norway during the Little Ice Age. *Journal of Arctic and Alpine Research*, 4, 131–8.

Grove, J. M. 1988: *The Little Ice Age*. London: Routledge.

Goudie, A. S. 1990: *The Landforms of England and Wales*. Oxford: Blackwell.

Goudie, A. S. 1993: Human influence on geomorphology. *Geomorphology*, 7, 37–59.

Goudie, A. S. 1995: *The Changing Earth: Rates of Geomorphological Processes*. Oxford: Blackwell.

Goudie, A. S. 2000: *The Human Impact*, 5th edn. Oxford: Blackwell.

Hall, A. P. and Griffiths, J. S. 1999: A possible failure mechanism for the

AD 1575 'Wonder Landslide'. *The East Midland Geographer*, 21/2–22/1, 92–105.

Hoskins, W. G. 1955: *The Making of the English Landscape*. London: Hodder and Stoughton.

Hutchins, J. 1803: *The History and Antiquities of the County of Dorset*, Vol. 2. J. Nichols and Sons.

Hutchinson, J. N. 1965: The stability of cliffs composed of soft rocks, with particular reference to the coasts of South East England. Unpublished Ph.D. thesis, University of Cambridge.

Hutchinson, J. N., Poole, C., Lambert, N. and Bromhead, E. N. 1985: Combined archaeological and geotechnic investigations of the Roman Fort at Lympne, Kent. *Britannia*, 16, 209–36.

Innes, J. L. 1983: Lichenometric dating of debris-flow deposits in the Scottish Highlands. *Earth Surface Processes and Landforms*, 8, 579–88.

Johnson, R. H. 1987: Dating of ancient, deep-seated landslides in temperate regions. In M. G. Anderson and K. S. Richards (eds), *Slope Stability*. Chichester: Wiley, 561–600.

Jones, D. K. C. 1991: Human occupance and the physical environment. In R. J. Johnson and V. Gardiner (eds), *The Changing Geography of the United Kingdom*. London: Routledge, 382–428.

Jones, D. K. C. 1999: Landsliding in the Midlands: a critical evaluation of the National Landslide Survey. *The East Midland Geographer*, 21/2–22/1, 106–25.

Jones, D. K. C. and Lee, E. M. 1994: *Landsliding in Great Britain*. London: HMSO.

Kates, R. W., Turner, B. L. and Clark, W. C. 1990: The great transformation. In B. L. Turner, W. C. Clark, R. W. Kates, J. F. Richards, J. T. Matthews and W. B. Meyer (eds), *The Earth as Transformed by Human Action*. Cambridge: Cambridge University Press, 1–17.

Lamb, H. H. 1982: *Climate, History and the Modern World*. London: Methuen.

Lamb, H. H. 1985: Climate and landscape in the British Isles. In S. R. J. Woodell (ed.), *The English Landscape: Past, Present and Future*. Oxford: Oxford University Press, 148–67.

Lamb, H. H. 1991: *Historic Storms of the North Sea, British Isles and Northwest Europe*. Cambridge: Cambridge University Press.

Lee, E. M. 1995: *The Occurrence and Significance of Erosion, Deposition and Flooding in Great Britain*. London: HMSO.

Lyell, C. 1835: *Principles of Geology*, Vol. III. London: Murray.

Morgan, R. P. C. 1985: Assessment of soil erosion in England and Wales. *Soil Use and Management* 1, 127–31.

Nir, D. 1983: *Man, a Geomorphological Agent: an Introduction to Anthropic Geomorphology*. Jerusalem: Keter.

Price, R. J. 1983: *Scotland's Environment During the Last 30 000 Years*. Edinburgh: Scottish Academic Press.

Prince, H. C. 1964: The origin of pits and depressions in Norfolk. *Geography*, 49, 15–32.

Rackham, O. 1985: Ancient woodland and hedges in England. In S. R. J. Woodell (ed.), *The English Landscape: Past, Present and Future*. Oxford: Oxford University Press, 68–105.

Rackham, O. 1986: *The History of the Countryside*. London: Dent.

Reid, C. 1884: Dust and soils. *Geological Magazine*, NS, 1, 165–9.

Roberts, N. 1998: *The Holocene*, 2nd edn. Oxford: Blackwell.

Robinson, D. N. 1969: Soil erosion by wind in Lincolnshire, March 1968. *East Midlands Geographer* 4, 351–62.

Sherlock, R. L. 1931: *Man's Influence on the Earth*. London: Thornton Butterworth.

Skempton, A. W. and Weeks, A. G. 1976: The Quaternary history of the Lower Greensand escarpment and Weald Clay vale near Sevenoaks, Kent. *Philosophical Transactions of the Royal Society of London*, A283, 493–526.

Stamp, L. D. 1964: *Man and the Land*. The New Naturalist No. 44. London: Collins.

Thompson, F. M. L. 1985: Towns, Industry, and the Victorian Landscape. In S. R. V. Woodell (ed.), *The English Landscape: Past, Present and Future*. Oxford: Oxford University Press, 168–87.

Wallwork, K. L. 1974: *Derelict Land*. Newton Abbott: David and Charles.

Werrity, A. 1997: Short-term changes in channel stability. In C. R. Thorn, R. D. Hey and M. D. Newson (eds), *Applied Fluvial Geomorphology for River Engineering and Management*. Chichester: Wiley, 47–65.

Werrity, A. and Brazier, V. 1994: Geomorphic sensitivity and the conservation of fluvial geomorphology SSSIs. In C. Stevens, J. E. Gordon, C. P. Green and M. G. Macklin (eds), *Conserving our Landscape: Proceedings of the Crewe Conference, 1992*. Peterborough: English Nature.

White, G. 1789: *The Natural History and Antiquities of Selborne*. London: White, Cochrane and Co.

Whitmore, T. M., Turner, B. L., Johnson, D. L., Kates, R. W. and Gottschang, T. R. 1990: Long-term population change. In B. L. Turner, W. C. Clark, R. W. Kates, J. T. Matthews and W. B. Meyer (eds), *The Earth as Transformed by Human Action*. Cambridge: Cambridge University Press, 26–39.

Wilson, S. J. and Cooke, R. U. 1980: Wind erosion. In M. J. Kirkby and R. P. C. Morgan (eds), *Soil Erosion*. Chichester: Wiley, 217–51.

Woodell, S. R. N. (ed.) 1985: *The English Landscape: Past, Present and Future*. Oxford: Oxford University Press.

Wright, T. 1668: *Philosophical Transactions of the Royal Society*, Vol. III, 37, 722–5.

Chapter 4

Valley-floor and Floodplain Processes

Barbara T. Rumsby

4.1   Introduction

British river basins encompass a diverse range of valley-floor and floodplain environments and are affected by many different processes. Factors such as hydroclimate, land use and human activities, catchment geology, glaciation history and catchment relief may impact on valley-floor and floodplain processes. A key division has long been recognized between river basins in the north and west of Britain and those in the south and east (Gregory, 1997), in terms of their catchment characteristics and Holocene and historic fluvial development. In the uplands of the north and west, basins are developed on older, more resistant geologies, are largely within the ice limits of the last glaciation and experience high annual rainfall totals. Channels are typically gravel-bedded, often through to lower reaches, with wandering or divided planforms, and flashy flow regimes. High stream power and ample sediment availability gives a higher potential for channel and floodplain change. Many upland basins have experienced net incision during the Holocene, and present-day channels are inset into, and confined by, late glacial and earlier Holocene terraces. In the lowlands of southern and eastern Britain, basins are developed on younger, less resistant geologies and experience lower annual rainfall totals. Channels carry a larger fine-grained component, have cohesive banks, relatively sinuous planforms and well-developed floodplains. The north-west/south-east division is also apparent in the duration and degree of human occupation of river basins, with many lowland catchments experiencing a longer history of development and intervention.

During the last 1000 years, valley floors and floodplains in British river basins have been subject to a range of changes associated with climatic variability and human activities. The scale and impact of human activities has increased greatly over the millennium, with an exponential increase in population, intensification of agriculture, industrialization and urbanization. Alongside these anthropogenic factors, there have been marked climatic fluctuations, especially around the Medieval Warm Period and the 'Little Ice Age' (see chapter 1 for a discussion of climate change).

Amid current concerns over the consequences of recent warming trends and enhanced variability of rainfall, and of human-induced environmental impacts, the recording and understanding of the response of fluvial systems to past environmental changes is increasingly viewed as vital in the context of developing scenarios for future response. Study of the Medieval Warm Period, for example, would appear to offer an analogue for the late twentieth century/early twenty-first century warming trend. However, a major issue that has not yet been fully solved is how to separate out climatic and anthropogenic signals from the alluvial sedimentary and morphological record (Macklin et al., 1992a). One approach is to choose a study area where only one set of factors has operated; for example, a basin with little or no human modification (e.g. Gurnell et al., 2000). However, in the densely populated regions of Britain and continental Europe, this often restricts investigations to small headwater basins, which may not be representative of the system as a whole. In any case, any future climatic changes will be superimposed on top of the myriad of anthropogenic activities and structures existing within a catchment. It is valley-floor and floodplain response to these combined and overlapping environmental changes that we need to understand, and this remains a major challenge. A number of recent publications have provided some insight into these issues over a longer-term, Holocene timescale (Macklin et al., 1992a; Macklin and Lewin, 1993, 1997) and from an archaeological perspective (Brown, 1997; Macklin 1999). This chapter provides an overview of the major factors affecting valley floors and floodplains in the last 1000 years, drawing on examples from various British river basins. It also outlines a number of recent developments that offer great potential for unravelling fluvial responses.

4.2   The Context: Environmental Changes Affecting Valley Floors and Floodplains in the Last 1000 Years

The patterns, rates and magnitudes of the major environmental changes affecting Britain in the last 1000 years and their geomorphological consequences have been reviewed in chapter 2. The factors most

Table 4.1 Anthropogenic influences in river basins in the last 1000 years, listing indirect (A) and direct (B) alterations, and their potential impact on water (Q_w), sediment (Q_s) and pollutant (P) supply to fluvial systems (based on Schumm, 1977; Warner, 1991)

	Q_w	Q_s	P
A. Catchment land use			
Forest clearance	+	+	
Afforestation	−	+	
Intensification of agricultural activity (grazing, arable)	+	+	+
Changes in farming practices (fertilisers, pesticides)			+
Land drainage	+		
Expansion of settlements, urbanization	+	−	+
Mining activities	+	+	+
Industrialization			+
B. Direct channel and floodplain modification			
Fish ponds		−	
Mills	−		
Flood prevention structures	+	−	
Dams and reservoirs	− or +	−	
Engineering for navigation			
Removal of riverbank trees		+	
Channelization and flow regulation	− or +	−	

relevant for, or having greatest impact upon, valleys floors and flood-plains are associated with anthropogenic activities and climate changes and are highlighted below.

4.2.1 Anthropogenic factors

Table 4.1 divides anthropogenic factors into activities affecting catchment land use, most of which do not set out to modify the fluvial system but do so inadvertently, and direct channel and floodplain modification. The potential impacts, in very broad terms, on water, sediment and pollutant supply to fluvial systems are indicated. The cumulative effect of human impacts on river basins throughout the Holocene is both widespread and significant. Prior to the last millennium, many British river basins had been subject to major land use alterations since the Neolithic and Bronze Age (up to 4000 years), including clearance of natural vegetation cover to provide agricultural land. Hillslopes and valley floors in lowland catchments were generally more significantly altered in terms of land use, with little woodland remaining by the medieval period, but with limited direct channel and floodplain modifications (Gregory, 1995). Upland basins, especially those in northern Britain, tended to be cleared later, from the late Roman period onwards, but with a significant proportion still reasonably well-wooded around AD 1000. Clearance of natural forest cover has been linked with enhanced soil erosion, and a range of studies have demonstrated marked influx of sediment to channels and floodplains consequent on the first major woodland clearance to take place in river basins (Macklin and Lewin, 1997). Agricultural innovations during the medieval period, including the introduction of ridge-and-furrow cultivation and deeper ploughing methods, led to an intensification of agricultural activity across Britain and many areas of remaining forest cover were cleared at this time. In the uplands of northern England the introduction of intensive local sheep farming, associated with tenth-century Viking settlement, has been linked to overgrazing in some areas and enhanced hillslope erosion (Harvey and Renwick, 1987). In the seventeenth to nineteenth centuries there was continued intensification of agricultural activity, with changes in farming techniques, extensive land drainage and improvement allowing higher production levels. Many seasonally flooded water meadows bordering low-gradient valley floors were embanked and drained at this time, including areas in the fens, the Somerset Levels and the Humber wetlands (Van de Noort and Ellis, 2000). Hillslopes in steeper upland basins have been subject to extensive moorland gripping in the mid–late twentieth century, with significant increases in drainage density and drainage network extension. The key impact of land drainage is to increase run-off from hillslopes and reduce the time to peak flow, hence producing flashier run-off regimes.

Other significant changes on floodplains and valley floors during the last 1000 years have been associated with the expansion of settlements, industrialization and mining activities. Floodplains and valley floors have always been favoured locations for human settlement and for location of transport routeways. The marked increase in population of Britain after the Industrial Revolution, and the increasing urbanization of that population, resulted in considerable floodplain development. This has lead to significant compartmentalization of floodplains (Lewin, 1983) and confinement of channels, with significant implications for planform change and flooding. Urbanization decreases the permeability of hillslopes and floodplains by covering surfaces with impervious tarmac and concrete, this in combination with construction of drainage conduits, results in decrease in sediment availability and an enhanced amount and rate of run-off (Leopold, 1990). Increased flood flows downstream of urban areas have been associated with stream channel enlargement, with a doubling of channel capacity (Gregory et al., 1992). However, many studies of impacts of urbanization have focused on relatively small catchments (<100 km^2), and it is likely that less significant changes would be expected in larger catchments if only a small proportion of the catchment was urbanized.

There is a long history of mining activity in many parts of Britain, associated with extraction of fossil fuels, heavy metals and aggregates, and the consequences for channel and floodplain development may be profound (Macklin and Lewin, 1997). Prior to the medieval period, mining tended to have a limited impact on the landscape, since exploration, extraction and processing methods were undertaken manually and were small-scale. However, from the medieval period onwards, increasing mechanization of extraction and processing and, in particular, the introduction of water power, resulted in more extensive implications. These include disruption of the hydrological system (hillslope vegetation disturbance and drainage modification), accelerated slope erosion and increased stream sediment loads (Macklin and Lewin, 1997). However, the extent to which mining activities adversely affect channel and floodplain locations depends on the location of extraction (valley sides, valley floor, within channel), and the degree to which waste products can be contained and prevented from entering surface or ground waters (Macklin and Lewin, 1997). In Britain, the impacts of heavy metal mining, especially the lead and zinc industries, are particularly well documented, and two types of response have been identified (Macklin and Lewin, 1989): active transformation and passive dispersal. The former usually results in a change from meandering to braided channel patterns with valley-floor aggradation, followed by incision and reversion to a single channel after cessation of mining. Although large-scale heavy metal mining, such as that associated with the Wealden iron industry in southern England, was initiated during the late medieval period, the most dramatic impacts are associated with developments after 1700, peaking in the mid-nineteenth century (lead) and early twentieth century (zinc). After the late nineteenth century, preventative legislation led to a reduction in discharge of mining waste into rivers, and by the mid-twentieth century large-scale heavy metal mining operations had ceased in many areas. However, a legacy of pollution exists in many catchments affected by former mining activities, associated with the dispersal and reworking of toxic mining wastes. Mining dispersal patterns in alluvial sediments have been widely studied in Britain, and fine-grained mine tailings are susceptible to wind and water erosion and may be moved significant distances downstream (Macklin et al., 1994).

The extent to which sediment and pollutants derived from catchment land-use changes are propagated to floodplains and valley floors depends on many factors that effect the degree of coupling or connectivity between hillslopes and channels, which is influenced by hillslope gradient, valley-floor and floodplain width and sediment calibre. In basins with steeper hillslopes and narrower valley floors, such as is the case in much of upland Britain, there is generally a more direct link, and the potential for materials to reach the main channel is greater. That notwithstanding, by end of the twentieth century many British floodplains and valley floors had been extensively modified by human activities and, especially in southern and lowland Britain, are essentially managed environments. Indeed, Brown and Quine (1999) suggested that in the British Isles there are no floodplains with natural vegetation, few without channel management and none without some flow management through reservoirs and land drainage. The upland–lowland divide is also clear, with the most intensive agricultural activities focusing on the higher-quality agricultural land in the south and east, while upland basins have experienced less intensive agriculture, being dominated by pastoral activities, but have been subject to extensive mining activities associated with mineral extraction.

4.2.2 Climatic variability

It is only within the last 300 years that systematic, instrumental meteorological records have become available, and information on climate trends prior to that is based on documentary evidence and proxy sources (Lamb, 1977, 1984; Bradley and Jones, 1995; Barber et al., 1999). Variability is the key characteristic of temperature and rainfall in the last 1000 years, although different trends and patterns are recorded in upland and lowland areas, and according to which climate proxy is used (see chapter 1 for a fuller discussion). Key factors of hydroclimatic relevance include annual precipitation totals, spatial and temporal trends in precipitation distribution and the proportion of precipitation that takes the form of as snowfall. Temperature trends are also important, both direct (evapotranspiration rates, freeze–thaw activity) and indirect (through the control of natural vegetation). The impact of climate change on fluvial systems is strongly moderated by catchment vegetation and land use, in particular by the effectiveness of surface cover in protecting hillslopes from erosion. However, vegetation cover itself, at least under natural conditions, is largely controlled by climate; hence strong feedbacks are possible, which act to augment or dampen response of fluvial systems.

4.3   Valley-floor and Floodplain Response

4.3.1   Evidence

Evidence of valley-floor and floodplain activity in the last 1000 years includes direct observation, documentary and cartographic sources, as well as analysis of valley-floor morphology and stratigraphy. Direct observations, including repeat surveys, are used to monitor change at specific locations over short periods, often no longer than 2–3 years, although some programmes have persisted for 1–2 decades (Knighton, 1998). Documentary sources include scientific reports and accounts, survey notes, personal diaries and parish records, and may provide specific, quantitative information on channel change, sediment deposition or degree of flood inundation, but more usually are qualitative in content. For many British river basins, there are adequate documentary sources available for the period since c. AD 1700, but records are patchier prior to that. Further, records are limited to those areas where humans were present to document events; hence the more sparsely populated uplands are less well covered. There are many caveats over the use of documentary records, as they are subject to bias and interpretation, and care must be taken to cross-check sources (Hooke and Kain, 1982). Maps and aerial photographs have been widely used to investigate channel planform change and lateral reworking of valley floors (e.g. Hooke and Redmond, 1989; Leys and Werritty, 1999; Rumsby et al., 2001), and afford a series of snapshots of the channel and floodplain over a period of time. The First Edition OS maps, dating back to the mid-nineteenth century, provide the first large-scale (1 : 10 560) systematic coverage across Britain. Prior to that, more patchy coverage is provided by sources such as estate maps and plans, tithe maps and military surveys, which are often of varying accuracy and at a range of scales. Particular care is required to determine the flow stage represented on a map, often a notional 'winter' stage, and the date at which fluvial features were last revised (which may not be the same as the survey date; see Hooke and Kain, 1982). Map-based sources only provide information on lateral channel change and do not allow vertical activity (aggradation or incision) to be determined. Good quality aerial photographs (APs) are available for much of Britain for the period since the mid-twentieth century, having been flown by the RAF in the 1940s and by the OS since the 1950s. The utility of APs for deriving information on floodplain and valley-floor processes depends on their resolution (flying height) and whether stereographic overlap is available. Often, APs are used alongside cartographic sources to extract information on river planform change and bar development over time (e.g. Gurnell et al., 1994). However, the key advantage of APs over maps is that it is possible, using photogrammetric techniques, to extract surface elevations and thus to examine vertical topographic changes. Photogrammetry requires sufficient ground control, sterographic overlap and photography taken with a metric (survey) camera whose calibration parameters are known. Up until recently, most photogrammetry was undertaken manually and was extremely time-consuming, thus limiting application to short sub-reaches and small areas, but the development of automated digital processing has considerably widened the potential for studying reach-scale morphological change (Lane et al., 1993). The other major source of evidence on floodplain and valley-floor processes is geomorphological – analysis of sediments and landforms (palaeochannels, terraces) on the valley floor – and provides a direct record of past river activity. Mapping, surveying and stratigraphic analysis is required to establish age relationships and the relative sequence of events. Preservation of geomorphological evidence is not contingent on a human presence to record events, and is not subject to human perception and interpretation of events: however, stratigraphy may be complex and difficult to interpret due to post-depositional modification, varying preservation potential and a lack of suitable dating control.

Various types of dating control are available to determine the age sequence of fluvial chronologies over the last 1000 years (table 4.2). Many documentary and cartographic sources are date specific, although care has to be taken to determine the date of survey, which is often different to the date of publication or writing. Ideally, information should be cross-checked against a range of sources, but this is not always possible or practical. Error bands are rarely cited for this type of dating control. In terms of dating sequences and landforms where documentary records are not available, archaeological, radiometric, dendrochronological and lichenometric methods may be applied. However, a range of potential problems and caveats limit their application to specific contexts or over restricted time periods. Many of the approaches date material incorporated within, buried under or overlying alluvial deposits, and hence do not directly date the fluvial event itself. Further, unless material is demonstrably *in situ* (e.g. a tree stump with penetrating roots), it may have been subject to reworking and would provide an age older than the fluvial event. Radiocarbon dating of 'modern' samples (post-industrialization – the last 300 years) may

Table 4.2 The applicable age ranges of various methods available to date valley-floor and floodplain landforms and sediments over the last 1000 years (solid lines). Dashed lines represent periods in which evidence is patchy or less secure

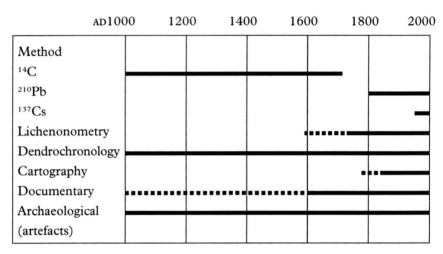

	AD1000	1200	1400	1600	1800	2000
Method						
¹⁴C						
²¹⁰Pb						
¹³⁷Cs						
Lichenonometry						
Dendrochronology						
Cartography						
Documentary						
Archaeological						
(artefacts)						

give anomalously old or young dates, due to the incorporation of geologically old carbon associated with fossil fuel combustion and new carbon associated with nuclear bomb testing, respectively. Other radiometric methods such as lead-210 (^{210}Pb) and caesium-137 (^{137}Cs) may provide dates covering the last 40–150 years (Walling and He, 1999). Lichenometry has been successfully applied to the dating of coarsegrained fluvial deposits over the past 250–300 years (Macklin et al., 1992b). This approach is limited to more remote areas and specific lichen species, as growth may be severely restricted by the high atmospheric pollution levels experienced in many British cities in the past 150 years. Hence, despite the wide range of methods potentially available for dating fluvial chronologies over the last 1000 years, certainly more than in the earlier Holocene, secure dating control is often problematic, especially in the gap between the application of radiocarbon (before *c.*300 BP) and the wider range of techniques available since *c.* AD 1800.

4.3.2 Modes of activity

Channels and floodplains adjust in various ways, which may be complex in space and time, and sudden changes can occur without a change in external conditions (Hooke, 1997). Broad categories of activity may be recognized, depending on the relative balance between water and sediment supply to fluvial systems and their implications for adjustment of slope, flow width and depth and median bed material diameter (Schumm, 1969). Where an excess of sediment supply exists, system will generally aggrade: alternatively, if there is an excess of stream power, channel degradation and enlargement will occur (bed incision and bank erosion). The style and rate of channel changes controls the morphology and sedimentology of floodplains, including the relative contribution of lateral and vertical accretion and continuity of deposition (Brown, 1996; Hooke, 1997).

More research has been undertaken on channel and floodplain evolution in meandering systems than on other types. Change in meandering channels is usually described in terms of morphological variables – meander wavelength, meander width and radius of curvature – and the main types of change involve meander migration and alterations in sinuosity (Hooke and Redmond, 1989). While numerous studies have attempted to relate such changes to particular causal factors, a major current research theme is the extent to which instability of meander characteristics is inherent in the system (Hooke, 1997). Morphological change in braided systems, and braid plain evolution, has received less attention and is not well understood. Braiding is usually attributed to situations in which there is an excess of sediment load and aggradation/deposition dominates, and is characteristic of streams with a high variability of discharge, a high sediment load, a wide valley floor, erodible bank materials and a steep gradient. Only one major braided system exists in Britain – the Spey and its tributaries, in Scotland – although many upland headwater streams display braided characteristics over short reaches and in response to specific events. There is some evidence to suggest that braiding was more widespread in the past in such systems (Passmore et al., 1993). The dynamism of morphological change in braided systems, with the potential for major changes to take place in single events, and variable response to similar events, makes braided rivers more difficult to analyse than meandering ones (Rumsby et al., 2001). There are also difficulties in defining morphological units due to the stage dependence of braiding indices: however, changes are closely related to the occurrence of competent flows (Werritty and Ferguson, 1980).

Styles of channel and modes of adjustment often show systematic differences in the downstream direction within a river basin. In this respect, Schumm's (1977) conceptual diagram of the fluvial system is useful, highlighting zones of supply, transfer and deposition in head-water, middle and lower reaches respectively. In gross terms, headwater

streams and hillslopes (Schumm's 'drainage basin') are zones of sediment supply, piedmont and middle reaches are zones of transfer, with
lower reaches and the ocean as major sinks of sediment. Investigations
in British river basins demonstrate a clear upland–lowland dichotomy
in terms of channel and floodplain processes and response. Upland
reaches are major sources of sediment, dominated by coarse sediment
deposition and episodic incision (Harvey et al., 1984; Macklin et al.,
1992b). Piedmont reaches, located at the upland/lowland divide,
appear to be most laterally mobile with high rates of planform change,
at least in the historical period (Hooke and Redmond, 1989). It is only
in the lower reaches that extensive floodplain development and significant depths of accretion occur, although the extent to which floodplain
development is currently taking place is debatable. Many lowland
reaches are essentially 'locked' into place through a process of relative
incision. Brown (1987) has attributed this as a response to supply of
predominately fine-grained sediment (sands, silts and clays) from hillslopes and valley floors in the mid–late Holocene, the cohesiveness of
which has allowed vertical banks to build up and channel enlargement
to occur. This has been termed the Stable Bed Aggrading Banks
(SBAB) model of floodplain development (Brown, 1987).

4.3.3 Chronology

In a review of British and European river basin response since
c.AD 1250, Rumsby and Macklin (1996) found an overall increase in
channel and floodplain activity over the period. Although this trend
may in part be attributed to better preservation and recording of more
recent evidence, there do seem to be consistent and widespread phases
of enhanced fluvial activity between 1250 and 1550, and particularly
between 1750 and 1900.

The Medieval Warm Period, AD 900–1250

Enhanced sediment supply is recorded in several upland basins at this
time, with alluvial fan and terrace aggradation reported in north-west
England (Harvey et al., 1981; Harvey and Renwick, 1987), the
Northern Pennines (Macklin et al., 1992b), the North York Moors
(Richards et al., 1987) and the Grampian Highlands (Ballantyne and
Whittington, 1999). In some cases, most notably in north-west England, fan aggradation has been linked to human-induced vegetation
changes, following Viking settlement (Harvey and Renwick, 1987). In
the lower reaches of several rivers draining upland areas, including the

Rheidol in mid-Wales (Macklin and Lewin, 1986) and the Dane in Cheshire (Hooke et al., 1990), significant alluviation of fine-grained units has been dated to the medieval period. This is generally attributed to accelerated soil erosion associated with intensification of agricultural activity and a rise in overbank sedimentation, although there is evidence of enhanced flooding in the eleventh century (Macklin, 1999).

AD 1250–1550

The beginning of this period, 1250–1350, saw a phase of increased flood frequency and magnitude recorded in many areas (Rumsby and Macklin, 1996), accompanied by a cooler, wetter climate. Valley-floor incision is recorded in headwater tributaries of the Tyne basin (Macklin et al., 1992c), and aggradation and braiding in reaches of the main channel downstream in the South Tyne valley (Passmore et al., 1993), indicating direct coupling of tributaries and main valley. Elsewhere, enhanced lateral shifting on reaches of the River Trent has been dated to between c.AD 1200 and 1600, in response to a series of large floods (Brown and Quine, 1999) and deposition of the contemporary flood-plain was initiated in Upper Wharfedale (Howard et al., 2000).

The Little Ice Age, AD 1550–1750

Relatively little upland fluvial activity has been dated to the coolest phase of the Little Ice Age, between 1550 and 1750, although enhanced freeze–thaw activity is likely to have increased the hillslope activity in upland areas. The relatively low rainfall totals at this time (Lamb, 1984) resulted in a low frequency of competent flows, and poor coupling between hillslopes and channels. Piedmont and lower reaches were more active, with lateral accretion and braiding being the dominant responses.

AD 1750–1900

From the late eighteenth century, a marked increase in fluvial response is apparent, particularly in upland basins, with widespread activity and significant vertical changes in many areas, commonly with channel incision in the late eighteenth century followed by aggradation or lateral accretion. Incision appears to coincide with an increase in flood frequency and magnitude (Rumsby and Macklin, 1996). Substantial deposition of fine-grained alluvium has been recorded on several piedmont floodplains downstream of major metal-mining areas (e.g. the River Swale at Catterick; Taylor et al., 2000).

AD 1900–2000

Major river regulation works in the late nineteenth century, and the construction of flood embankments, have limited the potential for significant lateral channel change in many lowland reaches, and rates of overbank vertical accretion are often low (Walling and He, 1999). In addition, there are sediment and water quality issues in many basins, associated with the widespread application of chemical fertilisers and pesticides on agricultural land and continued reworking of historical mining-age alluvium. Episodic incision has been recorded in a number of upland reaches, most notably in the mid–late twentieth century, in response to high-magnitude floods (Rumsby and Macklin, 1994).

Two major phases of enhanced fluvial activity are evident in the last 1000 years, centred on the eleventh century and the late eighteenth century. The first phase saw significant terrace and fan aggradation in upland reaches and floodplain alluviation in lowland reaches. In many basins, channel and floodplain change at this time has been linked with catchment land-use changes that accelerated hillslope and soil erosion, although there is also evidence of enhanced flooding, particularly in the eleventh century. The second major phase of activity in the late eighteenth century was marked by widespread incision, particularly in upland basins, and appears to be a response to a series of high-magnitude floods.

Differential response of upland and lowland reaches can be explained in terms of competence and sensitivity. Macklin (1999) suggests that the response of competence-limited coarse gravel bed rivers of northern and western Britain is closely related to climatic changes as reflected in flood frequency and magnitude fluctuations. The rivers of southern and eastern Britain, with their fine-grained sediment loads, are essentially supply limited and the effects of land use change are more significant. Furthermore, channel and floodplain evolution in many lowland systems since medieval times has been constrained by stabilization and embankment construction. Examples of channel change are developed in chapter 5.

4.4   Recent Developments

In the last decade, technological innovations and new applications of existing techniques have led to advances in interpreting floodplain and valley-floor stratigraphic sequences, and offer great potential for new levels of analysis and identification of patterns and causal linkages. Key areas include the resolution of environmental change records and dating control, geomorphological flood histories, remote sensing and numerical modelling approaches. These developments offer mechanisms for identifying form–process linkages and disentangling causal relationships for valley-floor and channel evolution over decadal to century timescales.

4.4.1 Resolution of Environmental Change Records and Dating Control

Advances in coring and sampling techniques have enabled valley-floor sequences to be sampled and subdivided into ever smaller increments, providing increased resolution and offering the potential to identify short-lived events in climate and fluvial records. However, the resolution capability of dating techniques is often not at the same level, or material suitable for dating is not present, and it is not always possible to establish the age of individual horizons. As discussed earlier, dating event horizons relating to the last 50–300 years is particularly problematic and several methods have been developed in recent years to fill this gap. These include the use of fallout radionuclides and heavy metal chemostratigraphy for dating fine-grained materials, and lichenometry for coarse-grained deposits.

Fallout radionuclides have been applied to examine soil erosion and redistribution and overbank sedimentation rates in many lowland British river basins. In a review of the use of fallout radionuclides to estimate rates of floodplain sedimentation, Walling and He (1999) present the results of studies on 25 lowland rivers, located mainly in south and west Britain. The study looked at two groups of sites, from which single or multiple cores were taken through fine-grained overbank sediments resulting from suspended sediment deposition during floods. The cores were dated on the basis of ^{137}Cs and unsupported ^{210}Pb measurements, and sedimentation rates calculated for the past 3–4 decades and the past 100 years, respectively (table 4.3).

The results show relatively low rates of annual floodplain accretion, ranging between 0.04 and 1.42 g cm^{-2} yr^{-1}, in lowland basins and higher rates from sites in catchments draining uplands. There is little difference in the time-averaged floodplain sedimentation rates in the last 40 years compared to the last 100 years, with a slight decrease predominating. This decrease is unexpected and requires further thought, given the significant expansion of erosion on agricultural land in the last 40 years. This apparent lack of sensitivity of floodplains to changes in sediment production might reflect the direct link between floodplain sedimentation and flood frequency, as opposed to hillslope

Table 4.3 Overbank sedimentation rates on British floodplains, showing average mean annual sedimentation rates based on ^{137}Cs and unsupported ^{210}Pb measurements (data taken from Walling and He, 1999)

River location	Mean annual sedimentation rates (g cm^{-2} a^{-1})		Trend[a]
Detailed study sites	*Past 40 years*	*Past 100 years*	
R. Culm, Silverton Mill	0.29	0.27	√
R. Severn, Buildwas	0.28	0.33	⇩
Warks. Avon, Eckington Bridge	0.09	0.08	↑
R. Rother, Shopham Bridge	0.22	0.20	√
Single core sites	*Past 33 years*	*Past 100 years*	
R. Ouse, York	0.95	1.04	√
R. Vyrnwy, Llanymynech	0.21	0.46	⇩
R. Severn, Atcham	1.22	1.42	⇩
R. Wye, Preston on Wye	0.15	0.28	⇩
R. Severn, Tewkesbury	0.86	0.95	√
Warwickshire Avon, Pershore	0.46	0.66	⇩
R. Usk, Usk	0.88	1.01	⇩
Bristol Avon, Langley Burrell	0.39	0.33	↑
R. Thames, Dorchester	0.51	0.64	⇩
R. Torridge, Great Torrington	0.70	0.93	⇩
R. Taw, Barnstaple	0.60	0.65	√
R. Tone, Bradford on Tone	0.56	0.43	↑
R. Exe, Stoke Cannon	0.45	0.42	√
R. Culm, Silverton	0.35	0.32	√
R. Axe, Colyton	0.51	0.40	↑
Dorset Stour, Spetisbury	0.04	0.04	√
R. Rother, Fittleworth	0.11	0.14	⇩
R. Arun, Billingshurst	0.39	0.48	⇩
R. Adur, Partridge Green	0.51	0.71	⇩
R. Medway, Penshurst	0.15	0.23	⇩
R. Start, Slapton	0.51	0.45	↑
Range	0.04–1.22	0.04–1.42	

[a] A difference of $> \pm 10\%$ was used as the threshold for identifying significant change in sedimentation rate (↑, increase; ⇩, decrease; √, no change).

sediment production (Brown and Quine, 1999). However, studies of contemporary suspended sediment yields suggest that a significant amount may bypass floodplains and be transported through the system (e.g. Lambert and Walling, 1987). A further possibility is that progressive floodplain sedimentation, without channel-bed aggradation or incision, will lead to relative incision and a reduction in overbank flooding.

A further method is based on tracing heavy metals released from catchment mining activities and relating concentrations in a number of river basins (Macklin, 1985). Analysis of heavy-metal chemostratigraphy may provide extremely tight age control in locations where accretion rates are rapid and the history of mining production is known in detail (e.g. Rumsby, 2000), but is of course limited to those basins that were subject to mining activities.

4.4.2   Geomorphological flood histories

One of the most significant developments in terms of investigating processes of longer-term floodplain and valley-floor evolution is recent advances in the identification and analysis of deposits and landforms associated with individual historical flood events and the construction of flood chronologies or histories (e.g. Rumsby, 1991). This has been particularly illuminating in upland basins in northern Britain, where few or no documentary flood records exist, but which are particularly susceptible to high-magnitude events. Small basins are sensitive to high-intensity rainfall events as the steep channel gradients and flashy run-off regimes produce floods with high unit stream powers and high rates of coarse sediment transport. Hence, in these environments valley-floor morphology and stratigraphy largely reflects extreme events, and investigation and dating of sequences allows reconstruction of the record of past large floods (Rumsby, 1991). The methodology was developed on headwater tributaries of the River South Tyne in the northern Pennines, where deposits associated with up to 21 flood events have been recognized, and lichenometric analysis has indicated that all but one of these has been deposited since the mid-eighteenth century (Rumsby, 1991; Macklin et al., 1992b). The reconstructed flood history demonstrated marked clustering of events, with periods of enhanced flood frequency and magnitude centered on the mid–late eighteenth century, the late nineteenth century and the mid-twentieth century. Subsequent work in the Yorkshire Dales (Merrett and Macklin, 1999) has confirmed the timing and wider significance of these periods, and the sensitivity of upland fluvial environments to changes in flood magnitude and frequency. Furthermore, the importance of high-magnitude floods in a formative sense in these systems means that they are susceptible to short-term changes in hydroclimate and sediment supply variations associated with catchment land use changes and may provide a detailed record fluvial response to environmental change (Rumsby and Macklin, 1994).

In some basins, and in specific events, where high suspended sediment loads are generated, the sedimentary record of flood events may be traced downstream to the main valley. In the Tyne system, fine-grained 'couplets' of flood sediment have been identified in floodplain and within-channel locations several tens of kilometres downstream (Macklin et al., 1992c). These downstream sequences often record more floods than their headwater counterparts, reflecting the lower entrainment thresholds of finer-grained material and the potential for sediment transfer and reworking in lower-magnitude events.

4.4.3   Remote sensing

Recent research in fluvial geomorphology has emphasized the direct monitoring of channel topography as a tool better to understand the interrelationship between river form and process, and in particular to estimate bedload transport rates and reach-scale sediment budgets (Lane et al., 1994; Brasington et al., 2000). Studies of river dynamics have often focused on planform changes and direct investigation of topographic change restricted to monitoring relatively widely spaced cross-sections over time. Within the last decade, developments in the spatial and spectral resolution of satellite remote sensing sensors and in the capability of airborne-based platforms have provided great opportunities for studying channel morphology and in-stream fluvial processes (Gilvear et al., 1999). The high spatial and spectral resolution and operational flexibility provided by airborne remote sensing is of particular relevance to British rivers, due to their relatively small size, and has opened up huge potential. A range of sensors are available, including survey cameras for aerial photography (high-resolution digital cameras are currently being developed), radar, laser (lidar) and spectral scanners (atm and casi). Coupling of remote sensing with GIS provides the opportunity to assemble and compare longer-term spatial data sets that aid quantification of channel and floodplain development.

To date, some of the most the most successful developments in terms of analysing three-dimensional morphological change over time come from the application of digital photogrammetry (Lane et al. 1993). For example, Lane et al. (1994) studied a dynamic proglacial stream using a combination of digital photogrammetry and conventional tachometric survey, to capture terrain data at very high densities. The distributed pattern of channel topographic change between daily melt events was assessed by directly comparing high-resolution DEMS of the same reach derived by repeated surveys. Application of a similar approach to analysis of archival aerial photography offers great potential for studying longer-term morphological development (Chandler and Brunsden, 1995), although the requirements of good ground control, the use of a metric camera with known calibration and on acceptable resolution/flying height mean that not all historical photography is suitable. One significant shortcoming to the photogrammetric approach is the difficulty of data acquisition from submerged areas of the channel. Westaway et al. (2000) have dealt with this by rectifying vertical imagery to take account of refraction at the water surface: however, in streams with high suspended sediment loads or coloured water, the channel bed may not be visible. A number of studies have reported some success at deriving subaqueous topography using multi- and hyperspectral image analysis, although only at relatively shallow depths (up to 90 cm), and in areas of low surface roughness (Winterbottom and Gilvear, 1997).

A complementary ground-based approach that offers rapid collection of surface and submerged topography is survey using the global positioning system (GPS) (Brasington et al., 2000). A combination of ground and air-based survey and data acquisition methods offer the potential for analysing morphological change at a range of nested time and space scales. Figure 4.1 illustrates this in the context of investigations of morphological and planimetric change on the River Feshie, Scotland, using the results of annual GPS survey, aerial photography analysis and geomorphological mapping (Brasington et al., 2000; Rumsby et al., 2000).

4.4.4 Modelling approaches

A range of fluvial and catchment-based numerical models have been developed in recent years, including computational fluid dynamics and two-dimensional finite element approaches applied to in-channel and floodplain flows (Bates et al., 1992). Some degree of success has been reported in terms of how well these approaches model elements of real fluvial systems. However, their complexity means they tend to have large computational requirements, which limits application to relatively restricted study reaches. In doing this, the models essentially fail to account for processes outside of the study reach (mass movement, hydrology and upstream sediment supply) and have difficulty in dealing

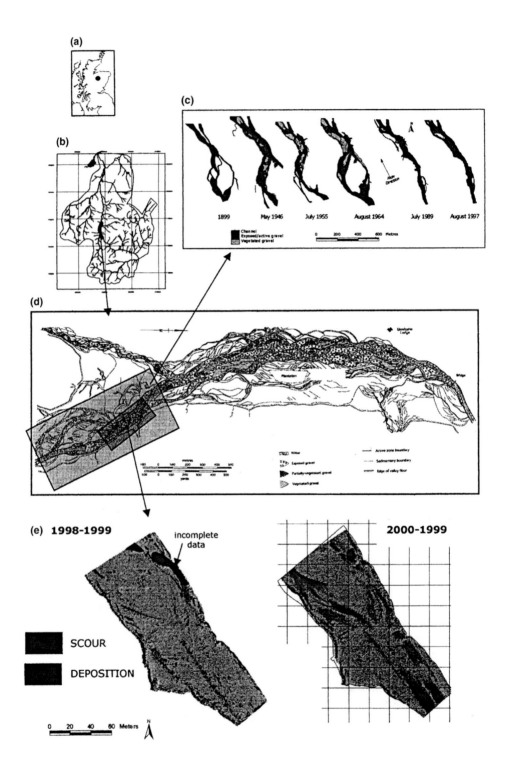

(a)

(b)

(c)

1899   May 1946   July 1955   August 1964   July 1989   August 1997

Channel
Exposed/active gravel
Vegetated gravel

0   200   400   600 Metres

Flow Direction

(d)

Lodge

Plantation

Bridge

Water
Exposed gravel
Partially-vegetated gravel
Vegetated gravel

Active zone boundary
Sedimentary boundary
Edge of valley floor

100   0   100   200   300   400   500
metres
100   0   100   200   300   400   500
yards

(e) **1998-1999**   **2000-1999**

incomplete data

SCOUR

DEPOSITION

0   20   40   60 Meters   N

with highly dynamic reaches (Coulthard et al., 1999). One type of numerical model that appears to offer potential for studying valley-floor and hillslope development at the catchment scale is a cellular automaton approach. Cellular models have recently begun to be applied to geomorphological contexts, including modelling channel and bar development in braided rivers (Murray and Paola, 1994) and impacts of environmental change in an upland catchment (Coulthard et al., 1999). Although cellular automaton models are based on a relatively simple set of rules, they may give rise to complex behaviour, and they appear to replicate patterns of change consistent with natural systems. In an interesting application, Coulthard et al. (1999) have modelled late Holocene channel, floodplain and hillslope evolution in Cam Gill Beck, a 4.2 km² tributary of the River Wharfe, in the Yorkshire Dales. A DEM of the catchment is divided into 1×1 m cells and each cell is assigned initial values for elevation, water discharge, water depth, drainage area and grain-size fractions. Hydrological, hydraulic, erosion and slope processes are represented by a series of equations applied to each cell and, after each iteration, cell values are updated. The effects of modelling 16 floods of around bankfull discharge appears to give realistic simulations of whole-catchment response, including non-linearity (sediment 'slugs'), slope–channel coupling and de-coupling, terrace, fan and boulder berm deposition and close similarity with observed forms in the study area. The results are promising, and highlight processes that need to be factored into the model, such as non-linear catchment inputs, including better representation of mass movements and, in order to quantify human impacts, the effect of vegetation on erosion and run-off.

Recent developments in the resolution of climate and flood records, dating methods, and the application of remote sensing and modelling approaches to investigation of fluvial response to environmental change offer great potential for enhanced understanding of form–process relationships and channel and floodplain evolution. However, there are still a number of unresolved issues, especially with respect to scaling: How do we scale up from process work to longer-term landscape evolution?

Figure 4.1 Linking time and space scales of channel and floodplain development, the River Feshie, Scotland (after Brasington et al., 2000; Rumsby et al., 2001). (a) Location map; (b) the River Feshie catchment; (c) historical channel and bar development, based on aerial photography and map analysis; (d) Holocene terraces; (e) recent morphological change on a 3 km reach, calculated from DEMs of difference based on GPS survey data.

4.5   Conclusions

There is a marked dichotomy in floodplain and valley processes in upland and lowland basins in the last 1000 years. Many lowland floodplains were subject to significant alluviation during the Medieval Warm Period, in response to enhanced sediment supply associated with catchment soil erosion. However, since then, channel migration and floodplain sedimentation has been increasingly restricted, by human-emplaced barriers (channelization and floodplain compartmentalization) and by the construction of natural silt–clay levees. Thus, many lowland reaches have shown little change, especially in the last 100 years. Upland floodplains and valley floors have been subject to less direct human interference, except in basins that experienced mining activities, and exhibit a close relationship with hydroclimatic variations, specifically changes in the timing and magnitude of geomorphologically effective floods. The phase of enhanced flooding in the late eighteenth century coincides with marked instability in a number of upland valleys, with high rates of channel-bed incision.

Research in a range of British river basins over the last five years or so has demonstrated increasing recognition of the non-linearity of fluvial response to environmental change. Brown and Quine (1999) provide a useful summary of some of the issues and stress the concept

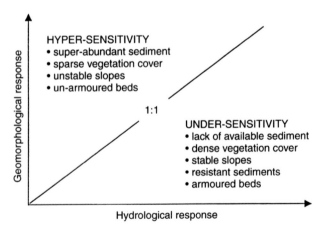

Figure 4.2 A schematic diagram illustrating factors influencing fluvial response to environmental change. Disproportionately large or small responses may occur due to hyper-sensitivity and under-sensitivity (after Brown and Quine, 1999).

of sensitivity, at the reach or catchment scale, as a crucial consideration when attempting to analyse causes of fluvial change. They suggest that sensitivity is conditioned by factors including sediment supply and calibre, hillslope and valley-floor vegetation cover and slope stability (figure 4.2). Reaches that are hyper-sensitive exhibit a disproportionately large geomorphic response from a small hydrological change, and those that are under-sensitive show a disproportionately small response (Brown and Quine, 1999). This distinction goes some way to explaining the divergent response of upland and lowland reaches to environmental changes in the last 1000 years. Many upland reaches fall into the hyper-sensitive category, with abundant sediment, unstable slopes and less dense vegetation cover. In contrast, the characteristics of many lowland valleys put then in the under-sensitive part of the diagram – supply limited, resistant banks and stable slopes.

REFERENCES

Ballantyne, C. K. and Whittington, G. 1999: Late Holocene floodplain incision and alluvial fan formation in the central Grampian Highlands, Scotland: chronology, environment and implications. *Journal of Quaternary Science*, 14, 651–71.

Barber, K. E., Battarbee, R. W., Brooks, S. J., Eglinton, G., Haworth, E. Y., Oldfield, F., Stevenson, A. C., Thompson, R., Appleby, P. G., Austin, W. E. N., Cameron, N. G., Ficken, K. J., Golding, P., Harkness, D. D., Holmes, J. A., Hutchinson, R., Lishman, J. P., Maddy, D., Pinder, L. C. V., Rose, N. L. and Stoneman, R. E. 1999: Proxy records of climate change in the UK over the last two millennia: documented change and sedimentary records from lakes and bogs. *Journal of the Geological Society, London*, 156, 369–80.

Bradley, R. S. and Jones, P. D. (eds) 1995: *Climate Since AD 1500*. London: Routledge.

Brasington, J., Rumsby, B. T. and McVey, R. A. 2000: Monitoring and modelling morphological change in a braided gravel-bed river using high resolution GPS-based survey. *Earth Surface Processes and Landforms*, 25, 973–90.

Bates, P. D., Anderson, M. G., Baird, L., Walling, D. E. and Simm, D. 1992: Modelling floodplain flows using a two-dimensional finite element model. *Earth Surface Processes and Landforms*, 17, 575–88.

Brown, A. G. 1987: Holocene floodplain sedimentation and channel response of the Lower River Severn, U. K. *Zeitschrift für Geomorphologie N. F.*, 31, 293–310.

Brown, A. G. 1997: *Alluvial Geoarchaeology: Floodplain Archaeology and Environmental Change*. Cambridge: Cambridge University Press.

Brown, A. G. 1996: Floodplain palaeoenvironments. In M. G. Anderson, D.

E. Walling and P. D. Bates (eds), *Floodplain Processes*. Chichester: Wiley, 95–138.

Brown, A. G. and Quine, T. A. 1999: Fluvial processes and environmental change: an overview. In A. G. Brown and T. A. Quine (eds), *Fluvial Processes and Environmental Change*. Chichester: Wiley, 1–27.

Chandler, J. H. and Brunsden, D. 1995. Steady state behaviour of the Black Ven mudslide: the application of archival analytical photogrammetry. *Earth Surface Processes and Landforms*, 20, 255–75.

Coulthard, T. J., Kirkby, M. J. and Macklin, M. G. 1999: Modelling the impacts of Holocene environmental change in an upland river catchment using a cellular automaton approach. In A. G. Brown and T. A. Quine (eds), *Fluvial Processes and Environmental Change*. Chichester: Wiley, 31–46.

Gilvear, D. J., Bryant, R. and Hardy, T. 1999: Remote sensing of channel morphology and in-stream fluvial processes. *Progress in Environmental Science*, 1, 257–84.

Gregory, K. J. 1995: Human activity and palaeohydrology. In K. J. Gregory, L. Starkel and V. R. Baker (eds), *Global Continental Palaeohydrology*, Chichester: Wiley, 151–72.

Gregory, K. J. 1997: An introduction to the fluvial geomorphology of Britain. In K. J. Gregory (ed.), *Fluvial Geomorphology of Great Britain*. London: Chapman and Hall.

Gregory, K. J., Davis, R. J. and Downs, P. W. 1992: Identification of river channel change due to urbanization. *Applied Geography*, 12, 299–318.

Gurnell, A. M., Downward, S. R. and Jones, R. 1994: Channel planform change on the River Dee meanders, 1876–1992. *Regulated Rivers*, 9, 187–204.

Gurnell, A. M., Edwards P. J., Petts G. E. and Ward J. V. 2000: A conceptual model for alpine proglacial river channel evolution under changing climatic conditions. *Catena*, 38, 223–42.

Harvey, A. M. and Renwick, W. H. 1987: Holocene alluvial fan and terrace formation in the Bowland Fells, northwest England. *Earth Surface Processes and Landforms*, 12, 249–57.

Harvey, A.M., Alexander, R. W. and James, P. A. 1984: Lichens, soil development and age of Holocene valley-floor landforms, Howgill Fells, Cumbria. *Geografiska Annaler*, 66A, 353–66.

Harvey, A. M., Oldfield, F., Baron, A. F. and Pearson, G. W. 1981: Dating of post-glacial landforms in the central Howgills. *Earth Surface Processes and Landforms*, 6, 401–12.

Hooke, J. M. 1997: Styles of channel change. In C. R. Thorne, R. D. Hey and M. D. Newson (eds), *Applied Fluvial Geomorphology for River Engineering and Management*. Chichester: Wiley, 237–68.

Hooke, J. M. and Kain, R. J. P. 1982: *Historical Changes in the Physical Environment*. London: Butterworths.

Hooke, J. M. and Redmond, C. E. 1989: River channel changes in England and Wales. *Journal of the Institution of Water and Environmental Managers*, 3, 328–35.

Hooke, J. M., Harvey, A. M., Miller, S. Y. and Redmond, C. E. 1990: The chronology and stratigraphy of the alluvial terraces of the River Dane valley, Cheshire, NW England. *Earth Surface Processes and Landforms*, 15, 717–37.

Howard, A. J., Macklin, M. G., Black, S. and Hudson-Edwards, K. 2000: Holocene river development and environmental change in Upper Wharfedale, Yorkshire Dales, England. *Journal of Quaternary Science*, 15, 239–52.

Knighton, D. 1998: *Fluvial Forms and Processes: a New Perspective*. London: Arnold.

Lamb, H. H. 1977: *Climate: Past, Present, Future*. London: Methuen.

Lamb, H. H. 1984: Climate in the last 100 years: natural climatic fluctuations and change. In H. Flohn and R. Fantechi (eds), *The Climate of Europe Past, Present and Future*. Dordrecht: Reidel, 25–65.

Lambert, C. P. and Walling, D. E. 1987: Floodplain sedimentation: a preliminary investigation of contemporary deposition within the lower reaches of the River Culm, Devon, UK. *Geografiska Annaler*, 69A, 47–59.

Lane, S. N., Chandler, J. H. and Richards, K. S. 1994: Developments in monitoring and terrain modelling small-scale river bed topography. *Earth Surface Processes and Landforms*, 19, 349–68.

Lane, S. N., Richards, K. S. and Chandler, J. H. 1993: Developments in photogrammetry; the geomorphological potential. *Progress in Physical Geography*, 17, 306–28.

Leopold, L. B. 1990: Lag times for small drainage basins. *Catena*, 18, 157–71.

Lewin, J. 1983: Changes of channel patterns and floodplains. In K. J. Gregory (ed.), *Background to Palaeohydrology*. Chichester: Wiley, 303–19.

Leys, K. F. and Werrity, A. 1999: River channel planform change: software for historical analysis. *Geomorphology*, 29, 107–20.

Macklin, M. G. 1985: Floodplain sedimentation in the upper Axe valley, Mendip, England. *Transactions of the Institute of British Geographers*, NS 10, 235–44.

Macklin, M. G. 1999: Holocene river environments in prehistoric Britain: human interaction and impact. *Quaternary Proceedings*, 7, 521–30.

Macklin, M. G. and Lewin, J. 1986: Terraced fills of Pleistocene and Holocene age in the Rheidol valley, Wales. *Journal of Quaternary Science*, 1, 21–34.

Macklin, M. G. and Lewin, J. 1989: Sediment transfer and transformation of an alluvial valley-floor: the River South Tyne, Northumbria, UK. *Earth Surface Processes and Landforms*, 14, 223–46.

Macklin, M. G. and Lewin, J. 1993: Holocene river alluviation in Britain. *Zeitschrift für Geomorphologie Supplementband*, 88, 109–22.

Macklin, M. G. and Lewin, J. 1997: Channel, floodplain and drainage basin response to environmental change. In C. R. Thorne, R. D. Hey and M. D. Newson (eds), *Applied Fluvial Geomorphology for River Engineering and Management*. Chichester: Wiley, 15–45.

Macklin, M. G., Passmore, D. G. and Rumsby, B. T. 1992a: Climatic and cultural signals in Holocene alluvial sequences: the Tyne basin. In S. Needham and M. G. Macklin (eds), *Alluvial Archaeology in Britain*, Oxbow Monograph 27. Oxford: Oxbow Press, 123–39.

Macklin, M. G., Rumsby, B. T. and Heap, T. 1992b: Flood alluviation and entrenchment: Holocene valley-floor development and transformation in the British uplands. *Geological Society of America, Bulletin*, 104, 631–43.

Macklin, M. G., Rumsby, B. T. and Newson, M. D. 1992c: Historical floods and vertical accretion of fine-grained alluvium in the lower Tyne valley, northeast England. In P. Billi, R. D. Hey, C. R. Thorne and P. Tacconi (eds), *Dynamics of Gravel-Bed Rivers*. Chichester; Wiley, 564–580.

Macklin, M. G., Ridgway, J., Passmore, D. G. and Rumsby, B. T. 1994: The use of overbank sediment for geochemical mapping and contamination assessment: results from selected English and Welsh floodplains. *Applied Geochemistry*, 9, 689–700.

Merrett, S. P. and Macklin, M. G. 1999: Historic river response to extreme flooding in the Yorkshire Dales, northern England. In A. G. Brown and T. A. Quine (eds), *Fluvial Processes and Environmental Change*. Chichester: Wiley, 345–60.

Murray, B. A. and Paola, C. 1994: A cellular model of braided rivers. *Nature*, 370, 54–7.

Passmore, D. G., Macklin, M. G., Brewer, P. A., Lewin, J., Rumsby, B. T. and Newson, M. D. 1993: Variability of late Holocene braiding in Britain. In J. L. Best and C. S. Bristow (eds), *Braided Rivers*. Geological Society Special Publication No. 75, 205–29.

Richards, K. S., Peters, N. R., Robertson-Rintoul, M. S. E. and Switsur, V. R. 1987: Recent valley-floor sediments in the North York Moors: evidence and interpretation. In V. Gardiner (ed.), *International Geomorphology Part I*. Chichester: Wiley, 869–83.

Rumsby, B. T. 1991: Flood frequency and magnitude estimates based on valley-floor morphology and floodplain sedimentary sequences: the Tyne basin, NE England. Unpublished Ph.D. thesis, University of Newcastle upon Tyne.

Rumsby, B. T. 2000: Vertical accretion rates in fluvial systems: a comparison of volumetric and depth-based estimates. *Earth Surface Processes and Landforms*, 25, 617–31.

Rumsby, B. T. and Macklin, M. G. 1994: Channel and floodplain response to recent abrupt climate change: the Tyne basin, northern England. *Earth Surface Processes and Landforms*, 19, 499–515.

Rumsby, B. T. and Macklin, M. G. 1996: River response to the last neoglacial (the 'Little Ice Age') in northern, western and central Europe. In J. Branson, A. G. Brown and K. J. Gregory (eds), *Global Continental Changes: the Context of Palaeohydrology*, Geological Society Special Publication No. 115, 217–33.

Rumsby, B. T., Brasington, J. and McVey, R. 2001: The potential for high resolution fluvial archives in braided rivers: quantifying historic reach-scale channel and floodplain development in the River Feshie, Scotland. In D. Maddy, M. G. Macklin and J. Woodward (eds), *River Basin Sediment Systems: Archives of Environmental Change*. Rotterdam: Balkema, 397–419.

Schumm, S. A. 1969: River metamorphosis. *Proceedings of the American Society of Engineers, Journal of the Hydraulics Division*, 95, 255–73.

Schumm, S. A. 1977: *The Fluvial System.* Chichester: Wiley.

Taylor, M. P., Macklin, M. G. and Hudson-Edwards, K. 2000: River sedimentation and fluvial response to Holocene environmental change in the Yorkshire Ouse basin, northern England, *The Holocene*, 10, 201–12.

Van de Noort, R. and Ellis, S. 2000: Wetland Heritage of the Hull Valley: an Archaeological Survey. Humber Wetlands Project, Kingston upon Hull.

Walling, D. E. and He, Q. 1999: Changing rates of overbank sedimentation on the floodplains of British rivers during the past 100 years. In A. G. Brown and T. A. Quine (eds), *Fluvial Processes and Environmental Change.* Chichester: Wiley, 207–22.

Warner, R. F. 1991: Impacts of environmental degradation on rivers, with some examples from the Hawkesbury–Nepean system. *Australian Geographer*, 22, 1–13.

Werritty, A. and Ferguson, R. I. 1980: Pattern changes in a Scottish braided river over 1, 30 and 200 years. In R. A. Cullingford, D. A. Davidson and J. Lewin (eds), *Timescales in Geomorphology.* Chichester: Wiley, 53–68.

Westaway, R. M., Lane, S. N. and Hicks, D. M. 2000: The development of an automated correction procedure for digital photogrammetry for the study of wide, shallow, gravel-bed rivers. *Earth Surface Processes and Landforms*, 25, 209–26.

Winterbottom, S. J. and Gilvear, D. 1997: Quantification of channel bed morphology in gravel-bed rivers using airborne multispectral imagery and aerial photography. *Regulated Rivers*, 13, 489–99.

Chapter 5

Fluvial Processes

Janet M. Hooke

5.1 Introduction

Over the last 1000 years major environmental changes, both natural and anthropogenic, have occurred which have affected rivers, directly and indirectly. The major phases of climate change provide the basic framework of driving forces for change, but superimposed on this are human impacts. To understand the effects of these changes and variations and their pattern over time, both the driving forces and the responses and their controls need to be examined. Fluvial processes and associated river morphology are obviously profoundly affected by climate, particularly precipitation; however, because of the necessity of water for human life and the value of rivers as means of transport, sources of power and food, and floodplains as locations for habitation, river valleys have a long history of human activity and impact. Indeed, many early settlements and early civilizations were located on rivers. Britain, with its long history of settlement, has thus been much affected, and the impact in the last millennium reflects the growing population, the increase in scale and scope of activities and the complex interaction with the environment. In a review of rural areas nearly a century ago, Lamplugh (1914) remarked: 'In the course of my field-work in the rural districts, I am constantly struck with the effect of human culture upon the streams. Hardly in any particular has Man in a settled country set his mark more conspicuously on the physical features of the land.'

Rivers are therefore locations of major human impact and of high sensitivity to climatic influences. They are now increasingly valued environmentally as some of the few remnants of 'natural' landscape and processes, so the passage to a new millennium is a fitting time to review the scale and amount of changes in fluvial processes and landforms, the causes of change and the impacts of human activity, and to evaluate the implications for the future of rivers and their management in the UK.

The historical timescale was much neglected by fluvial geomorphologists in Britain until the 1970s. There was a general assumption that rivers in Britain were stable, apart from a few upland rivers. In the 1970s much work was undertaken on the mechanisms and dynamics of processes and some work on morphological changes, particularly pattern, but mainly on the timescale of the last 150 years or so (see, e.g., Knighton, 1973; Mosley, 1975; Hooke, 1977; Lewin, 1978). Various papers demonstrated that processes were much more rapid and morphological changes much greater on many streams in Britain than previously perceived (Gregory, 1977). During the late 1970s and the 1980s there was a great increase in work on this topic, mainly focused on impacts of human activities, as these were assumed to be the cause of channel changes (see, e.g., Petts, 1984; Brookes, 1985; Gregory, 1987). However, by the end of the 1980s and the early 1990s work on chronologies of change over the past few centuries purported to show the very great influence of climate on river channels (see, e.g., Macklin and Lewin, 1993). At the close of the 1990s, the debate about climate versus humans was thus very vibrant. It is of immense importance in terms of the future implications of global warming and the management of our rivers. The purpose of this chapter is to review the nature and scale of human activities that have affected rivers in Britain and to assess their contribution to river changes, to discuss the processes of change in river channels, and to examine the spatial distribution of channel changes, exemplified by some contrasting case studies of degrees and causes of change. Since much more work has been done by fluvial geomorphologists on human impacts of the last 100–200 years, this chapter will focus on the earlier part of the last millennium. Other major conceptual questions or issues within fluvial geomorphology which are addressed in this chapter are the debates over catastrophic versus continuous changes, and the integration of short-term variability and longer phases of change.

5.2 Sources of Evidence

Morphological and sedimentary evidence are the standard sources on which geomorphological inferences of change are made. These have been widely used in studies of river channel change in the last millennium, and have been discussed in greater detail in chapter 4. Morphological evidence particularly comprises former river channel courses and river terraces. Sedimentological evidence has been used to infer processes, sediment supply and flux and incidence of floods. Both can indicate the size, state of morphology and position of the channels. They cover the whole time-span of the millennium, though obviously erosional activity destroys previous evidence and sedimentation may bury it. Archaeological evidence has now been much used in conjunction with fluvial investigations, and each is informing the other from their inferences (Needham and Macklin, 1992; Brown, 1997). For example, the work at Hemington on remains of medieval bridges across the River Trent (Salisbury, 1992; Brown and Quine, 1999) has provided valuable insights into fluvial morphology and processes. The value of the use of historical documentary evidence – provided that the requisite accuracy checks and corroboration are applied and information used appropriately – has long been advocated by Hooke and Kain (1982). The use of such sources is now much more common and is now aided by advances in technology and GIS. Maps have been most widely used, but can only reliably be extended back about 150 years for anything other than very qualitative evidence. Other graphical sources such as pictures can provide contributory evidence. Documentary sources such as manorial rolls and monastic records can extend farther back in the millennium and provide much evidence on activities, if rather less on the state and processes of rivers. Many of these historical sources have long been used by historians and historical geographers, but Lewin (1987), McEwen (1987) and Trimble (1998), as well as Hooke and Redmond (1989b), have discussed the use of historical sources in relation to river processes and channel changes.

5.3   The Causes, Chronology and Effects of Changes

The major influences or causes of change in fluvial systems are usually divided into categories of natural/human-induced change or as autogenic/allogenic change. Autogenic changes will be discussed in relation to the processes of change in the next section, but a key question in interpretation of changes in river courses is the extent to which these changes are inherent and the extent to which they are caused externally.

5.3.1   Climate

The basic patterns of climate variation over the last millennium have been summarized in table 1.3. Although the Medieval Warm Period and the Little Ice Age have long been accepted, it is interesting to note other evidence emerging of much less distinct major phases and greater short-term variations. A key question, which we are now beginning to amass enough evidence to address, is also the degree of synchronicity of climate, even within the small island of Britain.

It is generally assumed in fluvial geomorphology that the main influence of climate on river morphology is through the magnitude and frequency of flood events. Much work has been undertaken to construct the chronology of floods over the past few hundred years in several basins (see, e.g., McEwen, 1989; Rumsby and Macklin, 1994; Longfield and Macklin, 1999; Merrett and Macklin, 1999) and the relationship with climate has been examined. However, the question of the controlling, dominant or effective floods has long been debated, and it is now being increasingly appreciated that it is not only the statistical magnitude and frequency that are important but the sequencing and spacing of floods. The same flood magnitude can have differing impacts depending on the state of the channel (Huckleberry, 1994; Hooke, 1996). The influence of climate on run-off generation is a topic of much research and modelling, but the influence of other aspects of climate and their contribution to fluvial processes in channels has perhaps been neglected. Modern process studies have demonstrated the influence of frost action and moisture – for example, on bank erosion – but these results have not been used to infer the impact on processes in the past; for example, during the Little Ice Age, when considerable evidence does indicate a much higher incidence of cold conditions, such as the occurrence of ice on the River Severn at Coalport in the period 1789–1800 (Brown, 1997). While the overriding control is the incidence of floods, care must be taken on inferences of occurrence due to the partial preservation of flood evidence.

5.3.2   Human activities

Gregory (1995) has very effectively summarized the general chronology of human activities and fluvial effects for the past millennium (table 5.1), including both direct and indirect effects. The nature and timing of these impacts will now be discussed in more detail, concentrating on the direct effects. Sheail (1988) has also published a useful review of the history of river regulation in the UK. The major direct alterations of rivers include construction of mills and weirs, river regulation and channelization, construction of dams, alteration of drainage, and irrigation.

Table 5.1 River response to anthropogenic influences in the last 1000 years (from Gregory, 1995)

Period	Human activity	River response	Region
Medieval times	Forest clearance for agriculture	Minor rise in water table of lakes; increased flooding and alluviation; rapid soil erosion; channel floor built up	Southern England
	Intensification of agricultural activity	High rates of sediment influx; increased gullying on hillslopes; meander pattern replaced by braided channel; increased flooding	Northern England, Southern England
	Changes in farming techniques employed: ridge-and-furrow cultivation, deeper ploughing	Increased sedimentation rates; rapid surface run-off; increased erosion and flooding	Central England
	Shift to grassland for wool industry; regeneration of woodland	Decline in erosion and sedimentation	Central England
	Modifications to waterways: fishponds, mills, flood prevention, impoundments	Widening and shallowing of channel; increased ground and surface water retention; channel straightened	Southern England
	Mining activities	Waste products deposited in channel and on floodplain; increased overbank flows; fine sediments contain increased Pb concentrations	Nortnern England, Southern England
Seventeenth and eighteenth centuries	Cessation of mining activity; abandonment of settlements	Channel incision	Northern England
	Modifications to waterways 2: improvement for mining, engineering for navigation	Reduced load transported by river; increased dynamics of river pattern; lengths of channel abandoned	Southern England
	Land drainage	Shorter times to peak flow; greater magnitude floods; increased drainage density of river system	Central England, Northern England

Period	Activity	Effect	Region
	Mining activities	Higher concentrations of Fe, Mn, Pb, Na and K from waste mining material; increased amount of fine and coarse sediment delivered to channels; increased overbank flows	Northern England, Southern England
Eighteenth and nineteenth centuries	Changes in farming techniques employed: modern cereal farming practised, introduction of root crops	Increased erosion from farmland; marked irregularities in discharge; channel wider and straighter	Scotland
	Modifications to waterways: impoundment for mining, engineering for navigation	River width and depth increased; change in sinuosity; parts of course abandoned	Southern England
	Cessation of deposition of mining waste through legislation	Reworking of metal-rich alluvium in major floods; extensive dispersal downstream	Northern England
	Land drainage	Increased erosion from drainage ditches	Scotland
	Decline in agricultural activity; woodland regeneration – pine plantations	Little sedimentation occurred	Southern England
	Mining activities	Selective sedimentation of clay particles; significant deposition of metal-rich gravels; valley floor transformation; substantial reduction in W : D ratio; frequent flooding	Northern England, Southern England, Wales
Nineteenth to mid-twentieth centuries	Changes in farming techniques employed – cereal crop cultivation on hillslopes	Severe surface run-off and erosion; larger and more frequent floods; excessive sedimentation from overbank flows; increased W : D ratio of bankfull channel; valley accretion	Wales
	Land drainage	Renewed channel incision; stream power increased	Northern England
	Cessation of mining activities	Channel incision; reduced channel accumulation; decline in metal concentrations in sediments	Wales, Northern England

Mills and weirs

If one looks back over the early part of the millennium, it is particularly the high density of mills and the extent of activity that is very striking, although spatially variable. The Domesday Book records the presence of all mills for each county in 1086. At this time there were 5624 water mills in England south of the Trent. However, mills were much more abundant in southern England, with 318 mills in Hampshire at that time (Darby and Campbell, 1962; and see figure 5.1) but only 19 in Cheshire (Darby and Maxwell, 1962). By medieval times, almost every village had a mill. There was a density of a mill at intervals of, on average, less than 3 km on the Thames. The configuration of mills varied: some were directly on a stream, some using weirs, some – particularly later – using mill leats, and some having mill ponds created by small impoundments (Ellis, 1978; Brown, 1997). The overshot water wheel was introduced in 1743, which necessitated offtakes of water being farther upstream from the mill because of the height required.

Many weirs had fisheries associated with them, or fish traps were constructed in various ways to provide a major source of food in the medieval period. The Domesday Book records 21 fisheries in Notting-hamshire in 1086, for example. Much of the ownership and manage-ment of land and the operation of mills and fisheries in medieval times was by monasteries and manors, and the feudal system of labour had a profound influence on the extent to which labour intensive activities using rivers were possible. As mills became ever more numerous, so too did the problems associated with their use. The natural shallows became shallower as silt accumulated behind the mill weirs. The mills were less of a hazard if built on mill leats, but disputes about rights to water and diversions led to medieval laws. The mills and weirs were also major impediments to navigation, which was a major activity, and this also led to many disputes. One of the promoters of river navigation, John Taylor, complained that the owners of mills and weirs were responsible for the 'miserable strange abuses of the rivers' (Willan, 1964). The relationship between mills and river navigation has been the subject of much discussion recently among historical geographers (Jones, 2000; Langdon, 2000). Domesday mills were for corn but later milling extended to sawing, fulling, ore crushing, hammers, bellows and mine drainage. In around 1200 water-powered hammers were introduced to the iron industry. Hammer ponds were excavated in the Weald such that William Camden in the 1590s said there were furnaces on every side '. . . to which purpose divers Brookes in many places are brought to run in one channel and sundry meadows turned into pools

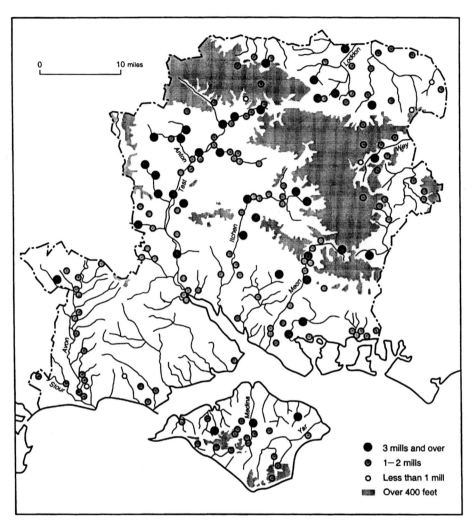

Figure 5.1   A map of mills in Hampshire at the time of the Domesday survey (after Darby and Campbell, 1962).

and waters' (Binnie, 1987). In the nineteenth century came the large expansion of textile mills so that, by 1788, 122 cotton spinning mills in Great Britain were motivated by water power, each with its own weir.

Early mills, prior to the Industrial Revolution, were arguably small-scale, but construction of leats were a major modification of channel pattern in many valleys. Many of these channel diversions still persist, even though use of the mills has disappeared and the mills themselves have decayed or been put to other uses. At the time of their operation and where structures are still intact, mills and weirs have the general effect of causing backwaters upstream and therefore a decrease in velocity but a greater propensity to flooding, and they usually cause localized erosion and widening immediately downstream of the structures, and shoaling and shallowing of waters downstream. Flow regimes would depend much on the mode and timing of operations of the mill. The hydrological and morphological impact of mills has received little systematic attention from fluvial geomorphologists. Likewise, in the rapid decline of water mills over the past two centuries, the effects of the decline or destruction of mills have not been examined systematically, although Johnson (1954) comments on the effects. Willan (1964) records that demolition of weirs on the Wye for navigation was opposed because they caused flooding that was essential, and he states that the demolition 'occasioned great Shoals and Other inconveniences'.

River navigation and channelization

River transport was very important in medieval times (Muir and Muir, 1986). In the Middle Ages rivers were extensively used for transport, with York at centre of the system of inland navigation and the Severn the waterway of the west: by the time of the death of Elizabeth I in 1603, the Thames had been used for transport for 500 years. However, various phases in the use of rivers and the efforts to make them navigable can be recognized. As Willan (1964) states:

> Before 1600 the attempts to improve navigation were spasmodic and the evidence of their success or failure is largely lacking. The period before Elizabeth was one in which legislation was passed mainly to preserve or restore navigation of rivers already naturally navigable; the period after her reign was taken up much more with attempts to make navigable new rivers. After 1750 the history of river transport is indivisibly bound up with that of canals.

How was river navigability established and maintained? Some dredging and canalization did take place in the Middle Ages, but much of the navigation was executed locally and thus few records exist. Acts of Parliament dealing with improvement of river navigation date from a statute of 1424 for the River Lea, which ordered the river to be scoured and made navigable by landowners on either side; later, they levied a rate for that purpose. It was the more widespread and systematic navigation involving larger-scale works that necessitated legislation and therefore leaves a much greater record: 'On the whole those powers [of

Navigation Acts] were extensive and included the general right to . . . scour the river and to cut its banks subject to the payment of compensation.' (Willan, 1964). Many cuts were made across meanders in the seventeenth and eighteenth centuries; for example, on the River Wey, the River Arun and the River Ouse in southern England. The Wey navigation to Guildford of 1653 improved the waterway with cuts across the necks of several meanders. On the River Ouse (Sussex), upstream of Lewes are cut-offs and levees created by canalization in the late eighteenth to early nineteenth centuries. Channel straightening downstream is a consequence of a nineteenth-century flood alleviation scheme. Levees stand above tidal mud flats that have been reclaimed since the sixteenth century (Burrin and Jones, 1991). Flash locks on rivers in Britain are recorded from as early as 1189 on the newly made Itchen navigation near Winchester. Pound locks were introduced on the Exeter canal in 1566, and seventeenth-century pound locks survive on the River Wey.

Disputes over the decrees of Commissioners of Sewers concerning the Medway, Dee and Wye at beginning of the seventeenth century show that decrees were issued ordering removal of weirs or tree stumps from rivers to facilitate the passage of boats. In 1600, floods along the banks of the Medway led to disputes as to whether damage was caused by weirs or iron works. Inhabitants declared that their weirs acted as fences across the river, increased fish and were not responsible for the floods. Owners of forges ascribed all damage to the weirs, which were 6–8 ft (1.8–2.4 m) high and caused considerable overflowing (Willan, 1964).

Willan (1964) states that, 'When the Acts speak of a river being "choked up and utterly unpassable" they appear usually to mean that the navigation was "impaired by sands, tree trunks, broken banks and weirs". The engineers then acted on the principle "Widen a Channel, and you weaken its Current; straighten, and you strengthen it: The first feeds and fills up the Channel, the last grinds and deepens it." ' He also states that in 1796 'The planting of willows or hedges was a recognised method of protecting banks.' Faggots and wood were also used. It is interesting that there have been recent calls to reinstate such practices in channel management in Britain (Environment Agency, 1999). Innumerable patents were taken out for machines for cleansing rivers, chiefly the Thames, in the early seventeenth century. In 1618, a patent was issued for a water plough for taking up sand and gravel. Later, there were complaints that the engines 'do much hurt to the river by taking up gravel from the firm ground and making great holes'. An Act was passed in 1790 to deepen the channel of the Severn between Stourport and Worcester by building projecting jetties, to increase the speed of the current and thus its scouring effect upon the river bottom (Brown, 1987).

There was a continuing history of river maintenance into the twentieth century, not so much for navigation as for flood control. There was a history of clearing and snagging on many rivers, and on some a practice of blowing up gravel bars was apparently used. Some clearance practices still persist or have only very recently been abandoned. Channelization increased through the nineteenth century, particularly in urban areas. However, the greatest increases and most profound alterations have occurred in the last 60 years. These have been well documented by Brookes and Gregory (1983) and show the very great extent and relatively high density of channelization and maintenance in the UK, particularly in lowland Britain. Channelization is still proceeding, if in somewhat modified form, with a tendency for softer engineering techniques. Brookes (1988) has demonstrated very clearly the effects of channelization, geomorphologically and ecologically, and both upstream and downstream.

Dams and reservoirs

The effects of dams been well researched, particularly by Petts (1984) and the history of dams in Britain has been documented by Binnie (1987). It is mostly the large, modern dams the impact of which has been researched. Earlier ones were smaller in scale and had lesser effects. The first reservoir recorded in the last millennium was built in 1189 near Winchester, to provide water for a 'flash lock' on the newly made Itchen navigation. A fish pond at Alresford *c.*1200 had a dam 7 m high and 70 m in length (Schnitter, 1994). The first large structure in Britain >15 m high was the Coombs dam, completed in 1787, and the first dam <30 m high was at Entwhistle in 1837. Several reservoirs were constructed in association with the canal building of the eighteenth century. The earliest conventional, post-Roman water-supply reservoir was constructed at Whinhill in 1796 to supply Greenock (Muir and Muir, 1986). A large increase in the number of dams took place in the nineteenth century, with the major water-supply reservoirs of the Victorian era. Numerous studies in Britain and around the world have shown that the major effects of dams are to alter the discharge regime, usually decreasing peak flows, and resulting in degradation and narrowing of channels downstream and the formation of a wedge of sediment upstream. The actual response depends much on the channel substrate, the size of the dam, the operation, the tributary configuration and the sediment supply, and responses can be complex (Petts, 1979). There are few documented cases of dam collapse in Britain, one of the few being the Dolgarrog disaster in North Wales in 1925 (Newson, 1989). Dam construction has declined much in recent years and, although it continues worldwide, there is now a backlash against dam construction after the detrimental effects have been realized. In North America there have been successful campaigns to alter the operation of dams, most famously with the Hoover dam and its effects on the Grand Canyon, and there are now even moves to remove dams from rivers (Graf, 1996). Given the low sediment yields in Britain, few large reservoirs have become infilled as yet, although minor ponds from early dates in the millennium have been obliterated.

Drainage

Major alterations of drainage networks and river courses have been made for the purposes of drainage, mainly in estuarine marshland and coastal boundaries, but also in river wetlands. The history of drainage of the Fens and Somerset levels has been well documented and is well known (Williams, 1970; Darby, 1983; Purseglove, 1988) and the transformation of other areas has been graphically illustrated (see, e.g., Wood, 1981). Direct alterations of networks have also taken place to effect upland and field drainage (Ovenden and Gregory, 1980). As well as digging of ditches, the development of clay pipes and underdrainage and their exemption from tax in 1826 was of profound influence in extending field drainage (Purseglove, 1988).

One aspect of drainage that had a profound effect on river courses in chalkland valleys in southern England was the formation of water meadows from the seventeenth century to the early twentieth century. Various mechanisms of distributing water were used. Fully developed water meadows were important in the seventeenth and eighteenth centuries, and were most popular on chalk lands because of the even temperature and high nutrients of the water. The mechanisms involved the use of leats and sluices on steeper valleys. In flatter valleys, flooding took place upwards by a dam on the stream and inundation upstream. Flooding downwards was more popular in Hampshire, where water was distributed by head and carriers. These were very extensive in the valleys of the Itchen, Avon, Test and Meon. Sheail (1971) has described in detail how these water meadows were operated and the effects of their decline.

Land use change

A range of land use changes and of changes in land practices have effects upon fluvial processes and on river channels indirectly. Most of these alterations in fluvial systems come about as responses to modifications of the run-off regime and of the sediment yield. Much research, both in Britain and elsewhere, has documented the effects of land use changes. In general, a change from forest cover to pasture to arable increases run-off and peak flows and increases sediment yields to water courses. The actual changes in land use have also been documented and much discussed both generally for Britain, and for regions and specific locations. These have shown that the type and timing of land use change is varied spatially. Much of the major Holocene deforestation of Britain took place much earlier than the last millennium, especially in southern England, but significant areas of Britain were still forested at the time of Domesday and have since undergone clearance. However, Harvey et al. (1981) have argued for the Howgill Fells that it was not deforestation that produced the major alluviation of the tenth century but an increase in grazing, causing destruction of turf cover. A major phase that was widespread in England and is well documented was the massive increase in arable land in the medieval period, with practices such as ridge-and-furrow. Many researchers attribute this land use as the cause of alluviation in the late-Saxon–early medieval period on many rivers. Whether the cessation of sedimentation and the subsequent phase of erosion associated with incision of the eighteenth century, identified on many rivers, was a result of decrease in arable land, adjustment of the rivers and sediment exhaustion, or was due to a change in climate associated with the Little Ice Age, or indeed combinations of these, is much debated.

Although the effects of land use on sediment loads and processes have been much discussed, the effects on river channels and the implications for some of the activities described above should be borne in mind. For example, small rivers flowing through woodland would probably have been much influenced by the formation and destruction of woody debris dams (Gregory and Davis, 1992) with their associated effects on erosion, sedimentation and channel form. On large rivers much woody debris would be transported in channels in floods, forming obstructions and snags. Again, that can still be seen on some channels in Britain with riparian woodland, and contemporary studies in Italy (Gurnell et al., 2000) serve as an analogue of likely past and more natural conditions. Periods of high sediment yield and alluviation would have led to the formation of shoals and bars in rivers and a much greater incidence of braiding, especially with much less protection of banks. During phases of incision then, there would have been slumping and calving of banks, and sedimentation in downstream zones and/or an increased delivery of sediment to the coast.

In the later part of the millennium, urbanization becomes a major land use change, with the rate and intensity increasing rapidly after the Industrial Revolution. The effects of urbanization have also been well studied by hydrologists and geomorphologists (see, e.g., Graf, 1975; Roberts, 1989; Gregory et al., 1992), the increased amounts of run-off and decreased lag times resulting in higher peak flows and lower baseflows generally. These in turn have resulted often in channel widening and deepening in reaches where channels can adjust. In many cases the rivers have been channelized and protected through urban areas, so the adjustment is limited, but the effects can be propagated both up and downstream (Brookes, 1988). Perhaps the situation at the time when urbanization had rapidly increased but urban drainage and channel control had been little implemented is worth considering. It is graphically portrayed by Engels, describing the River Irk in Manchester in the 1840s (in Newson, 1992):

> Above the bridge are tanneries, bone mills and gasworks, from which all drains and refuse find their way into the Irk, which receives further the contents of the neighbouring sewers and privies. It may be easily imagined, therefore, what sort of residue the stream deposits. Below the bridge you look upon the piles of debris, the refuse, filth, the offal from the courts on the steep left bank.

It was such conditions, of course, that instigated the modern history of pollution control.

5.4   Processes and Rates of Adjustment

Deliberate use of river channels as outlined above would have resulted in direct alteration of channel morphology and, in turn, of processes. This may have resulted in further modification or adjustment within those reaches, or the effects may have been propagated up- and downstream, particularly where reaches were constrained. Indirect alterations in catchments or climatic fluctuations resulting in run-off and sediment changes would be expected to result in channel response in the directions indicated by Schumm's (1969) metamorphosis model (which have been illustrated in table 4.1). The difficulty in analysing changes, and the challenge for prediction of changes and effects of activities, is that the multivariate nature of river channel systems leads to an indeterminancy in channel response. As Hey (1978), Gregory (1987) and many others have shown, channels may respond in channel cross-section form, in pattern or slope, or in a combination of these factors. Which will occur depends on the channel characteristics, particularly the channel boundary resistance. The degree to which channels will respond depends mainly on stream power, with lower stream power rivers and reaches being much less sensitive and showing much less adjustment; for example, to channelization (Brookes, 1987). Gregory (1987) has documented the degree of adjustment to various types of channel change.

The dominant process in any particular reach of channel at any particular time depends on the balance of energy compared with the sediment supply. With excess energy conditions then erosion will take place and a deficit will cause sedimentation, but the channel-forming processes will only take place in competent floods. Rumsby and Macklin (1996) have identified three modes of river activity – lateral accretion and overbank sedimentation, channel aggradation and channel incision. They have identified changes in rivers in 50 year periods and demonstrated the prevalence of vertical instability after 1700 (see also chapter 4). The greatest evidence for channel change is for channel pattern, mainly because of the use of maps for the past 150 years, and because of some preservation of traces of location in floodplains and terraces. One question of interest in relation to channel pattern is whether complete channel transformations (metamorphoses) took place, or only gradual changes of type of pattern, or only of degree within a pattern. Several authors have demonstrated variations in degree of braiding or proportions of occurrence of braiding over time (McEwen, 1989; Passmore et al., 1993) and several have attributed the higher amounts of braiding to periods of higher incidence of moderate–extreme magnitude floods, in turn, often attributed to climatic variations (Rumsby and Macklin, 1994). A phase in the late part of the nineteenth century is identified as one with relatively high amounts of braiding on several rivers, such as the Severn (Passmore et al., 1993; Hooke et al., 1994). In other cases, the occurrence of braiding has been attributed to increased sediment loads; for example, from mining (Macklin and Lewin, 1989). Harvey (1987) showed how the influx of an enormous amount of sediment produced from debris flows in an extreme thunderstorm transformed channels in the Howgill Fells. Over the two decades since that event, the channels have gradually adjusted back to nearly their prior form.

Looking at it in the opposite way, one would expect in periods of alluviation for there to be occurrence of shoaling and braiding. Recent phases of arable land use, supposedly markedly increasing slope erosion, have failed to produce such effects on a large scale, and even rates of floodplain sedimentation have shown very little response (Walling and He, 1999 – see table 4.3 for rates). By implication, this means that the amounts of alluviation represented by some of the medieval terrace deposits in northern England, notwithstanding that they are in upper parts of valleys and may have occurred by channel sedimentation rather than by overbank deposition, indicate massive amounts of sedimentation at that time, presumably with concomitant morphological characteristics. Again, most of the evidence relating to changes in meandering form is derived from the historical maps and aerial photographs of the past 150 years or so. Fragmentary evidence of channel lateral instability has been obtained from sedimentary and morphological evidence, and in some places detailed archaeological excavation has allowed some courses and positions of channels to be dated. Thus the nature and rate of changes has been reconstructed; for example, very notably at Hemington (Brown, 1997), where the evidence of three bridges across the Trent and of a Norman mill have indicated changes from single to multiple channels and back over time. Similar changes have taken place 20 km downstream (Salisbury et al., 1984) but with different timing (Brown and Quine, 1999). Detailed cartographic evidence allows the rate and type of change in meander courses to be examined. Over this period, very active piedmont reaches in Devon and northwestern England showed little sign of stabilizing towards an equilibrium meander form, as conventionally postulated, but instead show progressive evolution of the meanders throughout the period, with increasing sinuosity of the channel course, and extension and increased curvature of bends. Sinuosity cannot increase indefinitely and therefore Hooke (1995) has identified, in the most rapidly changing streams, that they have reached the stage of intersection and cut-off of bends, and sinuosity is now fluctuating (figure 5.2). Not all active rivers and all reaches show this behaviour, but the increased sinuosity is widespread and progressive throughout the period. This would seem to imply that, once meander development begins in such rivers, a high degree of autogenesis is present in the process. Rates of change have been found to accelerate as meanders increase in curvature (Hickin and Nanson, 1975; Hooke, 1997b). The changes have not been found to be associated with any particular forcing functions, such as major floods or changes in land use. It may, however, still be that there was some trigger to initiate these changes, since so many channels exhibit them, and the widespread occurrence of a phase of incision in the late eighteenth century has already been identified.

The various reconstructions of flood chronologies have identified certain outstanding extreme events on many rivers and some of these coincide; for example, the highest floods on the Trent, Upper Severn, Wye, Stour and Ouse occurred in 1794–5. Major floods took place in 1948, 1998 and 2000. The 'magnitude–frequency' question about the persistence of flood effects and the effectiveness of large events com-

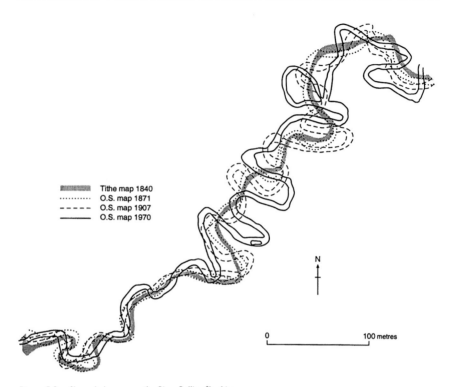

▓▓▓▓	Tithe map 1840
··········	O.S. map 1871
– – – –	O.S. map 1907
————	O.S. map 1970

N
↑

0 ——————— 100 metres

Figure 5.2   Channel changes on the River Bollin, Cheshire.

pared with more frequent ones over the long term has been addressed by some case studies in Britain, notably those of Anderson and Calver (1977), Carling (1986), Harvey (1986) and Newson (1980). Brown and Quine (1999), in analysing the evidence from Hemington, outlined above, suggest that the River Trent responded to a series of large floods between the eleventh and the fourteenth centuries, with major shifts of the channel. Yet the largest flood of all, in 1795, appeared to cause no morphological responses because, they surmise, by then the channel had become stabilized by incision and deposition of cohesive silt–clay banks and levees on top of the channel gravels. Although there are exceptions, the general pattern would seem to conform to Wolman and Gerson's (1978) ideas of much greater impact in the uplands and much lower sensitivity in the lowlands.

Rates of change have obviously varied both spatially and temporally over the past millennium. Of the most active rivers and maximum rates of change on meandering rivers, Hooke (1980) and Lawler (1987) have shown a general relation to discharge/catchment size worldwide. Evi-

Table 5.2   The ratio of depositional (d) to erosional (e) areas on the Upper River Severn in various periods

Period	D/E
1840–86	0.96
1886–1902	1.33
1902–47	1.04
1947–73	1.19
1973–84	1.90
1984–92	0.75

dence is increasing that adjustment of active rivers to alterations or fluctuations in regime, whether due to climate, land use or secular variations, may be quite rapid. Changes have been documented on the Severn (Thorne and Lewin, 1979; Higgs and Petts, 1988) and further work has shown phases of recent adjustment, probably associated with change in peak discharges (Hooke et al., 1994; and see table 5.2). Likewise, even in periods of a decade or so, the morphology, sedimentology and ecology of bars may change, and the dominance of erosion or deposition alter, probably in relation to flow incidence. For example, on the River Dane in Cheshire, the occurrence of unvegetated gravel bars has varied significantly over the past 20 years (Hooke, 1997a). Such rapid changes and fluctuations of channel conditions and even morphology beg the question of whether the recent period has been more variable than earlier periods, or whether the evidence of such variations becomes obliterated by larger-scale changes or by single events, or whether these small changes cumulate into apparent large-scale change. Understanding of this has implications not only for prediction of impacts of climate changes and land use change but also for the development and use of indicators and early warnings of change. This high sensitivity of some rivers must be borne in mind when we examine their millennial history. Have recent periods been more variable, or do we simply have more detail, and therefore should we infer the same degree of decadal changes farther back?

5.5   Spatial Patterns of Change

A crucial question as regards these changes over the last millennium is the degree of synchronicity of phases and behaviour or, as geographers, the complementary of spatial patterns and variability of change. To understand the patterns of change it is necessary to put together the distribution of forcing functions, both climate and human activities, and the sensitivity and ability of streams to adjust. If this is done, then a picture of spatial contrasts emerges. Questions of the spatial variability in climatic fluctuations have already been raised and are examined elsewhere. Some idea of the various indirect modifications has also been given. Spatial patterns of land use change still remain rather elusive to document in detail, especially at catchment scale and even for later parts of the millennium, let alone the medieval period. The spatial distribution of stream power gives some indication of the potential to respond to environmental changes. Stream power is the product of stream discharge and slope, and maxima have been shown by modelling to occur in the intermediate parts of basins (Knighton, 1999). The slopes are obviously higher in the uplands and greater in the upper than the lower parts of catchments, and discharge increases downstream in British catchments, so that the characteristics are spatially variable within and between catchments. Material tends to be coarser in the uplands and so requires higher competence for movement, and tends to be finer downstream, with cohesive, more resistant materials in the lowlands. Hooke and Redmond (1989a) have shown, for channel pattern changes over the past 150 years, that the greatest propensity to movement is in the piedmont zone, where not only is power usually highest but also resistance of material is lowest, river banks being mainly composed of erodible, sandy alluvium. It is difficult therefore to generalize on spatial patterns too much but, overall, we see a spatial variation in channel activity over the last millennium, with high natural channel change, lateral and vertical instability and high sensitivity to indirect changes in streams in the north and west, within and bordering the uplands and, generally, natural channel stability in the lowlands and the south, with low sensitivity to indirect changes but high amounts of direct channel interference. This produces a picture of strong natural stability of channels in the south, but a long history of channel usage and a direct alteration; whereas in the north there have been profound changes of vertical level and lateral stability of channels. Both Lewin (1987) and Brown (1987) are of the opinion that this spatial variability in channel stability is likely to have persisted for much of the Holocene. However, within these regional generalizations, individual catchments may still respond differently, and contrasts can be found between neighbouring catchments and even between reaches within a catchment.

5.6   Case Studies

These contrasts in environmental changes and channel behaviour are exemplified by the history over the last millennium of four different streams in England (figure 5.3).

5.6.1   Carling Gill, Howgill Fells

This is a small upland stream in Cumbria, which has experienced major changes and both vertical and horizontal instability, as documented by Harvey (1987) and Harvey et al. (1981, 1984), using a range of field evidence and dating methods. A major phase of aggradation which is now represented as a low terrace of coarse angular material, 1–2 m above the present valley floor, has been dated as post c.AD 940 and attributed to the introduction of sheep by the Vikings, and then expansion under monastic influence. This produced gullying on the hillslopes and aggradation in the valley floors. Incision to the present level was post this phase and pre-nineteenth century, as deduced from lichen dates on boulder bars. In the last 200 years the channel has switched in course but also incised. There are now active gullies contributing sediment in particular locations. Superimposed on this can be the occurrence of major storms; for example, an extreme event in 1982, which in neighbouring valleys produced major debris flows and increased sediment loading. This transformed channels for a period of some years, but these are now adjusted back to near previous conditions.

5.6.2   River Dane, Cheshire

This is a moderate-sized piedmont stream, representative of presently active meandering rivers (figure 5.4). The Dane valley and catchment was densely wooded in Domesday times (Hooke et al., 1990). Clearance took place in medieval times and was probably the cause of the major aggradation of fine material which took place, which increased the height of the river c.2 m above its previous level. In about the late eighteenth century major incision took place, leaving the aggraded material as a terrace. This change could be attributed to sediment decrease or channel adjustment, and/or could possibly be associated with the flood regime of Little Ice Age, or early industrialization upstream. The first mill is recorded in 1397, and the first silk mill in

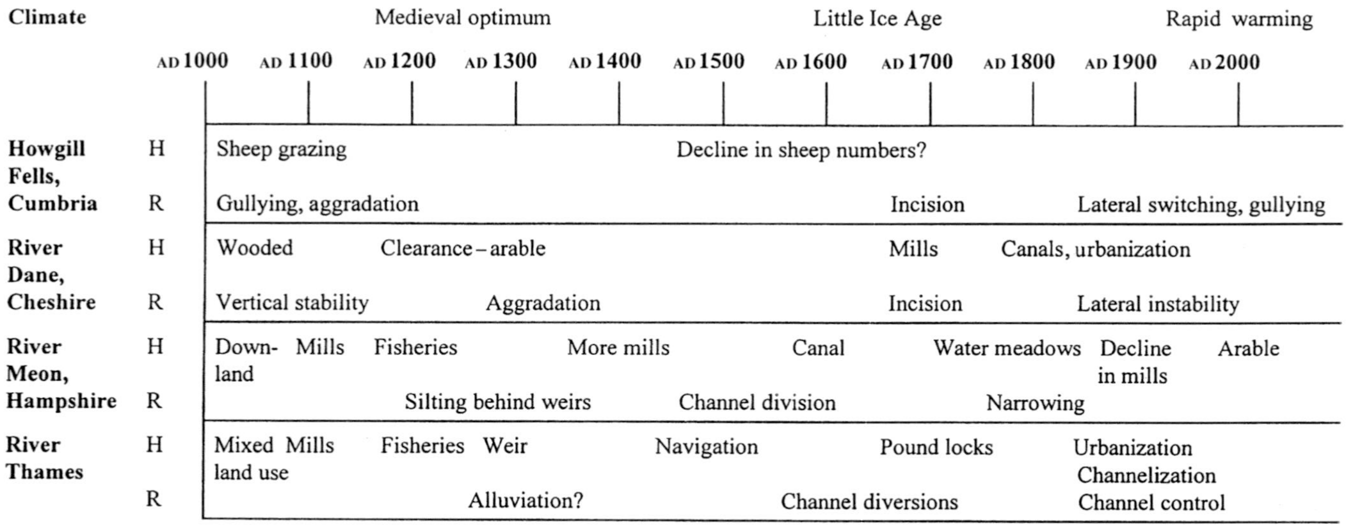

Climate		Medieval optimum					Little Ice Age			Rapid warming	
	AD1000	AD1100	AD1200	AD1300	AD1400	AD1500	AD1600	AD1700	AD1800	AD1900	AD2000
Howgill Fells, Cumbria H	Sheep grazing				Decline in sheep numbers?						
R	Gullying, aggradation							Incision		Lateral switching, gullying	
River Dane, Cheshire H		Wooded		Clearance–arable				Mills	Canals, urbanization		
R		Vertical stability		Aggradation				Incision		Lateral instability	
River Meon, Hampshire H		Down-land	Mills Fisheries	More mills			Canal		Water meadows	Decline in mills Arable	
R			Silting behind weirs		Channel division				Narrowing		
River Thames H		Mixed Mills land use	Fisheries Weir		Navigation			Pound locks		Urbanization Channelization Channel control	
R				Alluviation?			Channel diversions				

Figure 5.3 The chronology of changes on four contrasting streams. (H, human activity; R, river behaviour.)

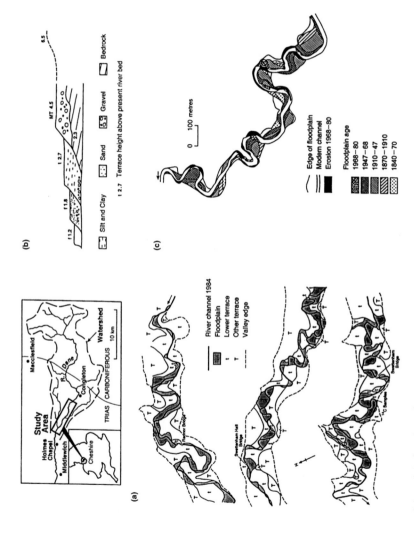

Figure 5.4 The morphology of the River Dane, Cheshire: (a) terraces and the floodplain in the valley floor; (b) a section through the terraces and floodplain; (c) the age of the modern floodplain from historical map evidence and aerial photographs.

1755. By 1817, 17 silk mills and five cotton factories were in operation upstream. The incision produced a relatively straight channel which, for the last 200 years, has been increasing in sinuosity, until recent attainment of the limit of sinuosity in some reaches produced cut-offs. In its central course, the river is still relatively natural and ecologically diverse. Some urbanization has taken place within the catchment.

5.6.3   River Meon, Hampshire

This is a small chalk stream, representative of a stable lowland river which has been subject to long human usage. It flows south from the South Downs, to enter the sea in the eastern Solent on the South Coast of England. Within a 6 km reach at the lower end, several historical features are present (figure 5.5). A mill site dates from Domesday and is still occupied by an old mill and a mill leat. The vertical stability of the floodplain since medieval times is illustrated by the so-called Anjou bridge. The Titchfield canal, which diverts water from the river and was built to facilitate navigation to Titchfield, was constructed in 1611. Water meadows were constructed in the seventeenth and eighteenth centuries. Henry Cort, inventor of the rolling mill for production of wrought iron, built a pond for his iron furnaces in 1784, which also involved diversion of tributaries from downstream to augment the water supply (Riley, 1971). Thus there is a history of usage and alteration of the channel, with the high proportion of the course that is modified apparent on figure 5.5, on what is naturally a low-energy, low-sensitivity channel.

5.6.4   River Thames

This, in its lower course, is a relatively large river, mainly in lowland, and a major navigable waterway subject to intense use and high urbanization. Maps of mills and fisheries in 1086 in Domesday times (Darby and Campbell, 1962) show the high density of activity even at that time. By medieval times there was a mill every 2 miles or less. The first weir was constructed in 1306. By the thirteenth and fourteenth centuries, fulling mills as well as corn mills were in operation. By the death of Elizabeth I, the river had been used for transport for 500 years. By the sixteenth century there were 23 weirs on 62 miles of river between Oxford and Maidenhead. In 1770, the navigation was entirely by flash locks, except for three pound locks near Oxford which were built in 1624–35. After 1771, eight pound locks were built between

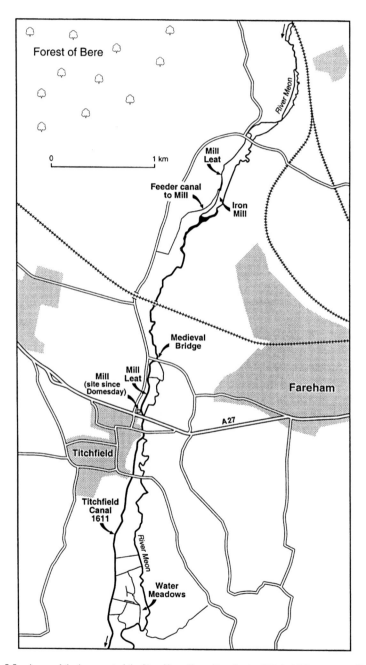

Figure 5.5   A map of the lower part of the River Meon, Hampshire, showing historical influences upon the present channel.

Reading and Maidenhead (Hadfield, 1966). There is some evidence to suggest some medieval alluviation in the Upper Thames (Limbrey, 1983; Robinson and Lambrick, 1984) and the Upper Thames was very difficult for navigation. Some channel change is also reported from archaeological evidence at Runnymede (Needham, 1992). Increasing urbanization and channelization have taken place throughout the nineteenth and twentieth centuries, and are still continuing. Some efforts have been made at channel restoration on tributaries in the urban area. Major flood relief schemes have been constructed, as at Maidenhead.

5.7   The Future

It seems unlikely that the spatial pattern of river activity will change fundamentally. Given the scenarios of global warming, the rivers of the uplands and piedmont seem likely to respond readily to shifts in river regime, possibly in complex ways. In the lowlands, it is likely that the major influence will still be increased pressure of population and occupancy of floodplain lands. Channelization and bank protection are still being implemented, in spite of policy statements to discourage this. Channel restoration is proceeding, but while local interests tend to favour protection, opportunities politically to implement such schemes are still fairly limited. Combinations of drought and flash floods may become characteristic of the run-off regime, but their effects will depend on land use practices and the degree of constraint of the channels. There has been a recent large increase in arable land in southern and eastern England, and it has been suggested that this change, in combination with the warming that is already apparent, could possibly be analogous to what happened in the medieval period. Given what we know at present about the impacts of changes, and seeing the extent to which we have already modified our river channels, we as a society must consider carefully how controlled a riverine environment we want in the future and how it may be sustained.

5.8   Conclusions

Changes in fluvial processes and river channels over the last millennium is obviously a vast topic, to which it is difficult to do justice in one chapter of a book. Nevertheless, some indication has been provided of the scope of human activities that have directly, and to some extent indirectly, affected river channels in Britain in that time. What emerges from this is the profound extent of human influence on our rivers and the large amount of activity even from early in this period. Opinion on whether the effects of that activity have overwhelmed the effects of climatic influence and the natural tendencies of the rivers varies amongst researchers. Evidence from a number of intensive studies around the country indicates that the degree of influence of these factors has also varied very considerably. Care must be taken not to overgeneralize and local variations must be recognized, but there are regional and locational contrasts. What is very striking is that, say in the sixteenth century, a river such as the Dane in Cheshire was 2 m above its present level and just emerging from a phase of rapid aggradation, with shoals and a highly mobile course, and still to experience a major incision and much subsequent lateral movement. So, even if there had been much human activity and many structures, we would expect to see little of them remaining. In contrast, by the same date, on the River Meon, a bridge which still survives had been constructed, as had mills, and a canal was about to be built, all at the present river level and all with much of the evidence surviving to the present day. Human impacts on rivers have tended to be cumulative: only on the most active rivers have traces of modifications been completely swept away or buried. Thus we are creating an increasingly constrained and anthropogenic environment, in which the flexibility of the river to adjust to changed conditions and the physical and ecological diversity of rivers is reduced. In spite of realizations that control strategies may not be sustainable, protection, channelization and flood alleviation schemes continue to be implemented. It should be accepted that erosion is not wrong, but can be a natural and necessary function for a river – and, with it, sedimentation. It seems difficult to envisage that our descendants will be able to write a comparable review of rivers at the end of the next millennium and not be dealing with a series of straightened ditch-like features in a very crowded land. On the question of climate or land use as causes of change, it would seem that the answer has to do with different causes at different times, in different places, and of combinations of causations and of feedback. Various research suggests that which cause is dominant depends on the sensitivity of the system, and on the relative magnitudes of the causes and the state of the system.

REFERENCES

Anderson, M. G. and Calver, A. 1977: On the persistence of landscape features formed by a large flood. *Transactions of the Institute of British Geographers*, NS, 2, 243–54.

Binnie, G. M. 1987: *Early Dam Builders in Britain*. London: Thomas Telford.

Brookes, A. 1985: River channelization: traditional engineering methods, physical consequences and alternative practices. *Progress in Physical Geography*, 9, 44–73.

Brookes, A. 1987: River channel adjustments downstream from channelisation works in England and Wales. *Earth Surface Processes and Landforms*, 12, 337–51.

Brookes, A. 1988: *Channelized Rivers.* Chichester: Wiley.

Brookes, A. and Gregory, K. J. 1983: An assessment of river channelization in England and Wales. *The Science of the Total Environment*, 27, 97–111.

Brown, A. G. 1987: Long-term sediment storage in the Severn and Wye catchments. In K. J. Gregory, J. Lewin and J. B. Thornes (eds), *Palaeohydrology in Practice: a River Basin Analysis.* Chichester: Wiley, 307–32.

Brown, A. G. 1997: *Alluvial Geoarchaeology: Floodplain Archaeology and Environmental Change.* Cambridge: Cambridge University Press.

Brown, A. G. and Quine, T. A. 1999: *Fluvial Processes and Environmental Change.* Chichester: Wiley.

Burrin, P. J. and Jones, D. K. C. 1991: Environmental processes and fluvial responses in a small temperate zone catchment: a case study of the Sussex Ouse Valley, southeast England. In L. Starkel, K. J. Gregory and J. B. Thornes (eds), *Temperate Palaeohydrology: Fluvial Processes in the Temperate Zone During the Last 15 000 years.* Chichester: Wiley, 217–52.

Carling, P. A. 1986: The Noon Hill flash floods. *Transactions of the Institute of British Geographers*, NS, 11, 105–18.

Darby, H. C. 1983: *The Changing Fenland.* Cambridge: Cambridge University Press.

Darby, H. C. and Campbell, E. M. J. 1962: *The Domesday Geography of South-East England.* Cambridge: Cambridge University Press.

Darby, H. C. and Maxwell, I. S. 1962: *The Domesday Geography of Northern England.* Cambridge: Cambridge University Press.

Ellis, M. 1978: *Water and Wind Mills in Hampshire and the Isle of Wight.* Southampton University Industrial Archaeology Group, Southampton.

Environment Agency 1999: *Waterway Bank Protection: a Guide to Erosion Assessment and Management.* Bristol: Environment Agency.

Graf, W. L. 1975: The impact of suburbanisation on fluvial geomorphology. *Water Resources Research*, 11, 690–92.

Graf, W. L. 1996: Geomorphology and policy for restoration of impounded American rivers: What is 'natural'?, In B. L. Rhoads and C. E. Thorn (eds), *The Scientific Nature of Geomorphology.* Proceedings of the 27th Binghamton Symposium in Geomorphology. Chichester: Wiley, 443–73.

Gregory, K. J. (ed.), 1977: *River Channel Changes.* Chichester: Wiley.

Gregory, K. J. 1987: River channels. In K. J. Gregory and D. E. Walling (eds), *Human Activity and Environmental Processes.* Chichester: Wiley, 207–35.

Gregory, K. J. 1995: Human activity and palaeohydrology. In K. J. Gregory, L. Starkel and V. R. Baker (eds), *Global Continental Palaeohydrology.* Chichester: Wiley, 151–71.

Gregory, K. J. and Davis, R. 1992: Coarse woody debris in stream channels in relation to river channel management in woodland areas. *Regulated Rivers,* 7, 117–36.

Gregory, K. J., Davis, R. J. and Downs, P. W. 1992: Identification of river channel change due to urbanization. *Applied Geography,* 12, 299–318.

Gurnell, A. M., Petts, G. E., Harris, N., Ward, J. V., Tockner, K., Edwards, P. J. and Kollmann, J. 2000: Large wood retention in river channels: the case of the Fiume Tagliamento, Italy. *Earth Surface Processes and Landforms,* 25, 255–75.

Hadfield, C. 1966: *British Canals: an Illustrated History,* 4th edn. Newton Abbot: David & Charles.

Harvey, A. M. 1986: Geomorphic effects of a 100-year storm in the Howgill Fells, NW England. *Zeitschrift für Geomorphologie,* 30, 71–91.

Harvey, A. M. 1987: Sediment supply to upland streams: influence on channel adjustment. In C. R. Thorne, J. C. Bathurst and R. D. Hey (eds), *Sediment Transport in Gravel-bed Rivers.* Chichester: Wiley, 121–50.

Harvey, A. M., Alexander, R. W. and James, P. A., 1984, Lichens, soil development and the age of Holocene valley floor landforms: Howgill Fells, Cumbria. *Geografiska Annaler,* 66A, 353–66.

Harvey, A. M., Oldfield, F. and Baron, A. F. 1981: Dating of post-glacial landforms in the Central Howgills. *Earth Surface Processes and Landforms,* 6, 401–12.

Hey, R. D. 1978: Determinate hydraulic geometry of river channels. *Journal of the Hydraulics Division, American Society of Civil Engineers,* 104, 869–85.

Hickin, E. J. and Nanson, G. C. 1975: The character of channel migration on the Beatton River, north-east British Columbia, Canada. *Bulletin of the Geological Society of America,* 86, 487–94.

Higgs, G. and Petts, G. 1988: Hydrological changes and river regulation in the UK. *Regulated Rivers: Research and Management,* 2, 349–68.

Hooke, J. M. 1977: The distribution and nature of changes in river channel patterns. In K. J. Gregory (ed.), *River Channel Changes.* Chichester: Wiley, 265–80.

Hooke, J. M. 1980: The magnitude and distribution of rates of river bank erosion. *Earth Surface Processes,* 5, 143–57.

Hooke, J. M. 1995: River channel adjustment to meander cutoffs on the River Bollin and River Dane, northwest England. *Geomorphology,* 14, 235–53.

Hooke, J. M. 1996: River responses to decadal-scale changes in discharge regime: the Gila River, SE Arizona. In J. Branson, A. Brown and K. J. Gregory (eds), *Global Continental Changes: the Context of Palaeohydrology.* London: The Geological Society, 191–204.

Hooke, J. M. 1997a: Interaction of sediment dynamics and channel morphology. In S. Wang, E. J. Langendoen and F. D. Shields (eds), *Management of Landscapes Disturbed by Channel Incision.* University of Mississippi, Oxford, Mississippi, 659–66.

Hooke, J. M. 1997b: Styles of channel change. In C. Thorne, R. Hey and M. Newson (eds), *Applied Fluvial Geomorphology for River Engineering and Management.* Chichester: Wiley, 237–68.

Hooke, J. M. and Kain, R. J. P. 1982: *Historical Change in the Physical Environment: a Guide to Sources and Techniques.* Sevenoaks: Butterworths.

Hooke, J. M. and Redmond, C. E. 1989a: River-channel changes in England and Wales. *Journal of the Institution of Water and Environmental Management,* 3, 328–35.

Hooke, J. M. and Redmond, C. E. 1989b: Use of cartographic sources for analysis of river channel change in Britain. In G. E. Petts (ed.), *Historical Changes on Large Alluvial European Rivers.* Chichester: Wiley, 79–93.

Hooke, J. M., Harvey, A. M., Miller, S. Y. and Redmond, C. E. 1990: The chronology and stratigraphy of the alluvial terraces of the River Dane Valley, Cheshire, NW England. *Earth Surface Processes and Landforms,* 15, 717–37.

Hooke, J. M., Horton, B. P., Moore, J. and Taylor, M. P. 1994: *Upper River Severn (Caersws) Channel Study,* Bangor, Countryside Council for Wales.

Huckleberry, G. 1994: Contrasting channel response to floods on the middle Gila River, Arizona. *Geology,* 22, 1083–86.

Johnson, E. A. G. 1954: Land drainage in England and Wales. *The Institution of Civil Engineers,* 6 April, 601–51.

Jones, E. T. 2000: River navigation in Medieval England. *Journal of Historical Geography,* 26, 60–75.

Knighton, A. D. 1973: Riverbank erosion in relation to streamflow conditions, River Bollin–Dean, Cheshire. *East Midland Geographer,* 6, 416–26.

Knighton, A. D. 1999: Downstream variation in stream power. *Geomorphology,* 29, 293–306.

Langdon, J. 2000: Inland water transport in Medieval England – the view from the mills: a response to Jones. *Journal of Historical Geography,* 26, 76–82.

Lamplugh, G. W. 1914: Physiographical notes. *Geographical Journal,* 43, 651–6.

Lawler, D. M. 1987: Climatic change over the last millenium in central Britain. In K. J. Gregory, J. Lewin and J. B. Thornes (eds), *Palaeohydrology in Practice: A River Basin Analysis.* Chichester: Wiley, 99–129.

Lewin, J. 1987: Historical river channel changes. In K. J. Gregory, J. Lewin and J. B. Thornes (eds), *Palaeohydrology in Practice: a River Basin Analysis.* Chichester: Wiley, 161–75.

Lewin, J. 1978: Meander development and floodplain sedimentation: a case study from mid-Wales. *Geological Journal,* 13, 25–36.

Limbrey, S. 1983: Archaeology and palaeohydrology. In K. J. Gregory (ed.), *Background to Palaeohydrology.* Chichester: Wiley, 189–212.

Longfield, S. A. and Macklin, M. G. 1999: The influence of recent environmental change on flooding and sediment fluxes in the Yorkshire Ouse basin. *Hydrological Processes,* 13, 1051–66.

Macklin, M. G. and Lewin, J. 1989: Sediment transfer and transformation of an alluvial valley floor: the River South Tyne, Northumbria, UK. *Earth Surface Processes and Landforms,* 14, 233–46.

Macklin, M. G. and Lewin, J. 1993: Holocene river alluviation in Britain. *Zeitschrift für Geomorpholgie Supplementband*, 88, 109–22.

McEwen, L. J. 1987: Sources for establishing a historic flood chronology (pre-1940) within Scottish river catchments. *Scottish Geographical Magazine*, 103(3), 132–40.

McEwen, L. J. 1989: River channel changes in response to flooding in the upper River Dee catchment, Aberdeenshire, over the last 200 years. In K. Beven and P. Carling (eds), *Floods: Hydrological, Sedimentological and Geomorphological Implications*. Chichester: Wiley, 219–38.

Merrett, S. P. and Macklin, M. G. 1999: Historic river response to extreme flooding in the Yorkshire Dales, northern England. In A. G. Brown and T. A. Quine (eds), *Fluvial Processes and Environmental Change*. Chichester: Wiley, 345–60.

Mosley, M. P. 1975: Channel changes on the River Bollin, Cheshire, 1872–1973. *East Midlands Geographer*, 6, 185–99.

Muir, R. and Muir, N. 1986: *Rivers of Britain*. Exeter: Webb and Bower.

Needham, S. 1992: Holocene alluviation and interstratified settlement evidence in the Thames valley at Runnymede Bridge. In S. Needham and M. G. Macklin, *Alluvial Archaeology in Britain*. Oxford: Oxbow, Monograph 27, 249–60.

Needham, S. and Macklin, M. 1992: *Alluvial Archaeology in Britain*. Oxford: Oxbow, Monograph 27.

Newson, M. 1980: The geomorphological effectiveness of floods – a contribution stimulated by two recent events in mid-Wales. *Earth Surface Processes and Landforms*, 5, 1–16.

Newson, M. 1989: Flood effectiveness in river basins: progress in Britain in a decade of drought. In K. Beven and P. A. Carling (eds), *Floods: Hydrological, Sedimentological and Geomorphological Implications*. Chichester, Wiley, 151–69.

Newson, M. 1992: *Land, Water and Development: River Basin Systems and Their Sustainable Management*. London: Routledge.

Ovenden, J. C. and Gregory, K. J. 1980: The permanence of stream networks in Britain. *Earth Surface Processes*, 5, 47–60.

Passmore, D. G., Macklin, M. G., Brewer, P. A., Lewin J., Rumsby, B. T. and Newson, M. D. 1993: Variability of late Holocene braiding in Britain. In J. L. Best and C. S. Bristow (eds), *Braided Rivers*. London: The Geological Society, 205–29.

Petts, G. E. 1979: Complex response of river channel morphology to reservoir construction. *Progress in Physical Geography*, 3, 329–62.

Petts, G. E. 1984: *Impounded Rivers: Perspectives for Ecological Management*. Chichester, Wiley.

Purseglove, J. 1988: *Taming the Flood: a History and Natural History of Rivers and Wetlands*. Oxford: Oxford University Press.

Riley, R. C. 1971: Henry Cort at Funtley, Hampshire. *Industrial Archaeology*, 8(1), 69–76.

Roberts, C. R. 1989: Flood frequency and urban-induced channel change:

some British examples. In K. J. Beven and P. A. Carling (eds), *Floods: Hydrological, Sedimentological and Geomorphological Implications*. Chichester: Wiley.

Robinson, M. A. and Lambrick, G. H. 1984: Holocene alluviation and hydrology in the upper Thames basin. *Nature*, 308, 809–14.

Rumsby, B. T. and Macklin, M. G. 1994: Channel and floodplain response to recent abrupt climate change: the Tyne Basin, Northern England. *Earth Surface Processes and Landforms*, 19, 499–515.

Rumsby, B. T. and Macklin, M. G. 1996: River response to the last neoglacial (the 'Little Ice Age') in northern, western and central Europe. In J. Branson, A. G. Brown and K. J. Gregory (eds), *Global Continental Changes: the Context of Palaeohydrology*. London: The Geological Society, 217–33.

Salisbury, C. R. 1992: The archaeological evidence for palaeochannels in the Trent Valley. In S. Needham and M. G. Macklin (eds), *Alluvial Archaeology in Britain*, Oxford: Oxbow, Monograph 27, 155–62.

Salisbury, C. R., Whitley, P. J., Litton, C. D. and Fox, J. L. 1984: Flandrian courses of the River Trent at Colwick, Nottingham. *Mercian Geologist*, 9(4), 189–207.

Schnitter, N. J. 1994: *A History of Dams: the Useful Pyramids*. Amsterdam: Balkema.

Schumm, S. A. 1969: River metamorphosis. *Journal of the Hydraulics Division, American Society of Civil Engineers*, 95, 255–73.

Sheail, J. 1971: The formation and maintenance of water-meadows in Hampshire, England. *Biological Conservation*, 3(2), 101–6.

Sheail, J. 1988: River regulations in the United Kingdom: an historical perspective. *Regulated Rivers*, 2, 221–32.

Thorne, C. R. and Lewin, J. 1979: Bank erosion, bed material movement and planform development in a meandering river. In D. D. Rhodes and G. P. Williams (eds), *Adjustments in the Fluvial System*. Dubuque, Iowa: Kendall/ Hunt, 117–37.

Trimble, S. W. 1998: Dating fluvial processes from historical data and artifacts. *Catena*, 31, 283–304.

Walling, D. E. and He, Q. 1999: Changing rates of overbank sedimentation on the floodplains of British rivers during the past 100 years. In A. G. Brown and T. A. Quine (eds), *Fluvial Processes and Environmental Change*. Chichester: Wiley, 207–22.

Willan, T. S. 1964: *River Navigation in England 1600–1750*, 2nd edn. London: Frank Cass.

Williams, M. 1970: *The Draining of the Somerset Levels*. Cambridge: Cambridge University Press.

Wolman, M. G. and Gerson, R. 1978: Relative scales of time and effectiveness of climate in watershed geomorphology. *Earth Surface Processes and Landforms*, 3, 189–208.

Wood, T. R. 1981: River management. In J. Lewin (ed.), *British Rivers*. London: George Allen & Unwin, 170–95.

Chapter 6

Estuaries and Coasts: Morphological Adjustments and Process Domains

E. Mark Lee

6.1 Introduction

The coastline has been the most dynamic element of the British landscape over the last 1000 years. Its varied form reflects the complex interaction between forcing processes – predominantly waves, but also tides, winds, and estuary flows – with the shoreline and nearshore morphology and materials, leading to the establishment of zones of high energy (wave convergence) and low energy (wave divergence). These zones are interlinked through sediment (clay, silt, sand and gravel) exchanges to form coastal cells, comprising both offshore and onshore elements. Simple cells comprise an arrangement of sediment source areas (e.g. eroding cliffs and the sea bed), areas where sediment is moved by coastal processes and sediment sinks (e.g. beaches, estuaries or offshore sinks). Along a particular stretch of coast there may be a series of such cells, often operating at different scales.

The energy arriving at the coast is considerably larger and more variable than to land-based systems. The 50 year storm event on the South Wales coast, for example, produces two orders of magnitude more energy than that produced by the 50 year flood event on the River Severn (Pethick, 2000). The combination of variable energy inputs and mobile sediment lead to morphological adjustments, ranging from beach profile changes over the course of a single storm to long-term changes in response to factors such as relative sea-level rise. A number of modes of change can be recognized (Pethick, 1992a):

1   *Episodic, progressive change*, where no recovery or retrogressive change occurs. For example, cliff erosion involves sequences of

episodic landslide events that result in loss of clifftop land. Cliffs do not recover by advancing forward.

2 *Periodic or cyclical change*, where the landform responds to an event (e.g. a storm or large flood) by altering the morphology, but then gradually recovers (either partially or completely) its original pre-event form. For example, variations in wave energy can cause major changes to beach form, with sediment being eroded from the upper shore and transported offshore under high wave energy conditions and brought onshore in lower-energy conditions, leading to beach accumulation (figure 6.1(a)). Thus, there may be an envelope of beach positions varying between steeper gradients associated with periods of beach deposition and flatter profiles associated with periods of beach erosion.

The availability of sediment for exchange or transfer between different 'stores' is critical to cyclical change. Considering a beach–sand dune system, sand may be deposited on the upper beach during low wave energy events, providing a source of sediment for aeolian transport and foredune growth. During high-energy events, the beach will flatten and the foredunes will erode. This provides an additional source of sand that is transported seaward and deposited on the lower beach.

Figure 6.1(b) illustrates cyclical behaviour for a saltmarsh, with vertical erosion occurring during storm events. Recovery of the marsh begins once the storm has passed, since the lower marsh surface results in lower wave shear stress, because of the deeper water depths and higher accretion rates. In the example shown in figure 6.1(b), the recovery from a 1 in 30 year storm takes place over a five year period (Pethick, 1992b). The mudflat–saltmarsh system behaves in a similar way to the beach–dune system described earlier. High-energy events lead to saltmarsh erosion, which supplies sediment to the fronting mudflats. Post-storm recovery involves mudflat lowering and saltmarsh accretion (figure 6.1(c)).

3 *Secular change*, where there is a longer-term trend in the evolution of coastal systems. On the coast, the Holocene sea-level rise and changes in sediment availability have initiated important long-term trends in landform development. During the early Holocene rapidly rising relative sea level resulted in onshore movement of large volumes of sand and gravel on to the coast and, in Scotland, this is associated with periods of beach and dune formation (Hansom, 1998). However, after around 6500 BP, relative sea level began to fall on parts of the coast, because of the increasing significance of isostatic uplift. The resultant sediment deficit had two main effects on coastal evolution in Scotland:

(a)

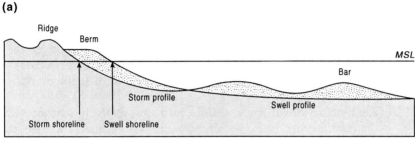

(b)

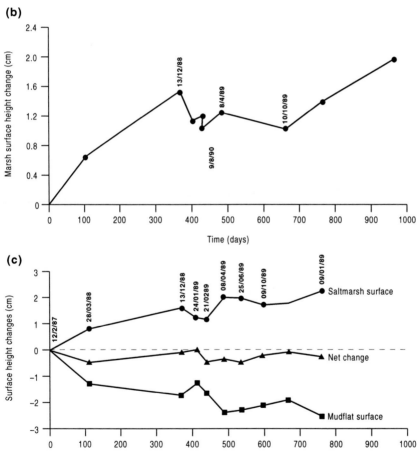

Figure 6.1 Models of coastal change. (a) A simple beach profile change model. Monitored changes at Dengie Peninsula, Essex. (b) The erosional response of a saltmarsh surface to a storm event on 20 December 1988, and its subsequent recovery. (c) Changes in mudflat and saltmarsh surfaces over a two-year period (from Pethick, 1992a).

- Erosion of the beach and back-beach dune systems, as sediments were removed from these areas into deeper water in order to maintain a nearshore gradient capable of dissipating wave energy. The widespread trend of beach erosion and eroding dune systems throughout the Western and Northern Isles is believed to be the consequence of the long-term sediment deficit (Hansom, 1998).
- Reorganization of the coastline into progressively smaller sediment transport cells, as the emergence of headlands on the retreating shoreline created barriers to longshore transport. These cells have tended to evolve subsequently by internal reorganization, erosion and deposition.

This chapter concentrates on providing an overview of the more dramatic progressive and secular changes that have occurred within the last 1000 years. However, rather than simply include a catalogue of change around the coast, a framework will be presented (and illustrated with examples) that will show how these changes can be viewed as part of the longer-term adjustments that have probably occurred throughout the current interglacial period. In addition, man has had a major impact on coastal landforms, especially over the last 500 years. This theme will also be explored within the chapter.

6.2   Domains of Coastal Change

A number of authors have suggested that coastal landforms have originated and evolved in response to changes in a number of fundamental controls (Forbes et al., 1995; Orford et al., 1996; Brunsden and Lee, in press). These controls include relative sea-level rise, geology and topography, sediment supply and wave climate and tidal range. Over the Holocene timescale, the fundamental controls appear to have been sea-level and sediment supply, with geology and inheritance providing the framework within which these controls operate.

Sea level rose rapidly during the early Holocene, with a marked deceleration in the rate around 5000–6000 years ago (see figure 2.2). There is believed to be a clear relationship between the rate of relative sea-level rise and the volume of sediment released into the coastal zone (Hansom, 1998; Jennings et al., 1998). This is based on the assumption that the principal source of beach-building sediment was a veneer of glacial and periglacial material landward of the retreating shoreline. As sea level rose, this material was pushed onshore. During periods of faster sea-level rise, beach development was probably controlled by the interaction of sea-level rise, sediment supply and longshore transport. However, as the rate of sea-level rise decreased, sediment supply and longshore transport declined substantially. Given the history of sea-level rise around Britain, it is possible that sediment supply to the beaches has declined substantially since around 5000–6000 BP, providing a long-term control on beach development and the dominant influence of sediment scarcity. Sea level has been effectively stationary for the last 2000 years and seabed sources are now largely exhausted or immobile. New coarse sediments tend to be limited to the longshore transport of materials released by cliff recession and fluvial inputs.

The coastal changes over the last 1000 years can be described in terms of a number of process domains that reflect variations in the fundamental controls. Following the previous work on shingle beaches (Forbes et al., 1995; Orford et al., 1996) and cliffs (Brunsden and Lee, in press), five domains are proposed:

1   *Initiation*. This domain follows the initial reorganization of coastal sediment around the coast as a distinctive suite of landforms and renewal of marine erosion at the foot of the coastal slope. Generally, this has coincided with postglacial sea-level rise, although man's intervention has promoted the development of new forms in some areas.

2   *Consolidation*. This domain is associated with the development of a characteristic form (a set of forms that develop and persist through time, although individual elements are evolving) and the establishment of a balance between process and form. For some landforms, such as cliffs and estuaries, this condition can be regarded as reflecting steady state behaviour, characterized by maintenance of form and a balance over time in the sediment budget. For others, the characteristic form may be slowly changing over time; that is, in dynamic equilibrium.

3   *Adjustment*. This domain marks the development of a new characteristic form in response to a significant change in one or more of the fundamental controls.

4   *Breakdown*. This domain marks the onset of unstable behaviour as the landform enters a period of reorganization into a number of discrete sub-units, marked by changing patterns of erosion and accretion. Ultimately, this may lead to the loss of a particular landform.

5   *Abandonment*. This phase follows the cessation of erosion and accretion processes along the shoreline, either in response to sea-level fall (as at the end of the Ipswichian interglacial), the growth of coastal landforms such as beaches, saltmarshes or spits that protect

a cliff foot, such as the Hadleigh cliffs, Essex and the Hythe–
Lympne escarpment, Kent (Hutchinson and Gostelow, 1976;
Hutchinson et al., 1985; Bromhead et al., 1998), or due to coast
protection.

The different phases do not represent a natural evolutionary
sequence of a deterministic trend, with breakdown or abandonment
the logical consequence of consolidation. Rather, they are a reflection
of changing controls over time. The fact that forms may re-occur
suggests that the range of interacting conditions lead to inherently
stable responses. Thus, a cliff characteristic form might be viewed as
the most likely stable configuration for the prevailing conditions.
Changes in the controls may or may not lead to the development of a
new characteristic form. However, in other circumstances landforms
may switch domains at different times. Carter and Orford (1993)
suggested that the changes between a shingle barrier prone to overtopping (consolidation domain) and one prone to overwashing (breakdown
domain) may be catastrophic, while the return path may be more
gradual.

The transition between domains may involve the exceedance of
critical thresholds and abrupt change. Thus, different domains may be
viewed as part of an overall metastable equilibrium that contains the
potential for catastrophic change (Jennings and Orford, 1999). An
alternative view of the interlinked *consolidation–adjustment–breakdown*
domains is one of continuous change and increased instability over
time; that is, change is inherent, and ongoing change with little stability
might occur.

6.2.1 Initiation

The overwhelming majority of coastal landforms were initiated in
response to the Holocene sea-level rise and, hence, were in existence
prior to the start of the last millennium. Although major changes have
occurred, such as the growth of spits and dune systems at estuary
mouths, it is argued that these reflect adjustments to pre-existing
landform assemblages in response to factors such as land reclamation,
port and harbour development and changes in sediment supply.

The exception to this pattern is the apparently recent initiation of
recession along some clifflines. The fall in sea level during the Deven-
sian glaciation stranded an Ipswichian cliffline, which was then subject
to slope degradation processes, under periglacial or glacial conditions.
Extensive spreads of landslide debris, head and screes accumulated at

the foot of the Ipswichian cliffline. Evidence of these features remains at East Cliff, Lyme Regis and at other sites around the south-west coast, including Porlock in Somerset (Jennings et al., 1998) and on the Start–Bolt coast in South Devon (Mottershead, 1997). As sea level rose after the glaciation, it would have gradually reached the degraded landslide deposits and solifluction sheets that would have been trimmed back, resulting in the creation of a new cliffline. However, degraded periglacial slopes still occur directly behind some sand or shingle beaches, where there appears to have been no or very little marine erosion of the foot of these slopes through the Holocene. Projection of the degraded landslide slope profile seaward at East Cliff, Lyme Regis and on the Gurnard to Cowes and Osborne clifflines (Isle of Wight) suggests that little cliff-foot retreat has occurred and that recession was only initiated within the last 1000 years (High-Point Rendel, 2000; Halcrow Group Ltd, 2000).

6.2.2 Consolidation

Fringing beach/shore platform–cliff systems provide a good example of landform assemblages that have probably maintained a characteristic form throughout the last millennium while undergoing almost continual change. The cliff recession process involves a repetitive sequence of events (a 'cycle') driven by factors such as debris removal from the foreshore and periods of high groundwater levels (Brunsden and Lee, in press). Four main stages of activity can be recognized: pre-failure movements, failure, post-failure movements and the reactivation of the displaced material. Each stage in the 'cycle' involves a different set of controlling factors and their relative significance will vary between different cliffs. Table 6.1 presents a summary of recent cliff recession rates for England and Wales, although the extent to which these rates are representative of longer-term rates over the last 1000 years is uncertain.

The Holderness cliffs on the East Yorkshire coast, for example, have retreated some 2 km over the last millennium, during which time 26 villages listed in the Domesday Book have been lost. Recession involves complex patterns of events of varying magnitude. This is well illustrated by measurements by the local authority at a series of 71 marker posts since 1951 (figure 6.2). Analysis of this data indicates that the observed average annual recession rate of 1.82 m comprises an average of 2.7 failures of 0.68 m depth per year (Pethick, 1996a). The periodicity in peak recession rates is explained in terms of the southerly movement of sand along the beach, driven by northeasterly waves. This movement

Table 6.1 A selection of reported average recession rates around the coast of England and Wales (after High Point Rendel, in press)

Site	Erosion (m yr⁻¹)	Period	Source
Aberarth, Dyfed	0.12	1880–1970	Jones and Williams (1991)
Llanon, Dyfed	0.25		Jones and Williams (1991)
Morfa, Gwynedd	0.08		Jones and Williams (1991)
Llantwit, South Glamorgan	0.43		Williams et al. (1991)
Ogmore–Barry, South Glamorgan	0.07	1977–85	Williams and Davies (1987)
Blue Anchor Bay, Somerset	0.2		Williams et al. (1991)
Downderry, Cornwall	0.11	1845–1966	Sims and Ternan (1988)
St Marys Bay, Torbay	1.03	1946–75	Derbyshire et al. (1975)
Bindon, East Devon	0.1	1904–58	Pitts (1983)
Charton Bay, East Devon	0.25	1905–58	Pitts (1983)
Black Ven	3.14	1958–88	Chandler (1989), Bray (1996)
Stonebarrow, Dorset	0.5	1887–1964	Brunsden and Jones (1980)
West Bay (W), Dorset	0.37	1887–1962	Jolliffe (1979), Bray (1996)
West Bay (E), Dorset	0.03	1902–62	Bray (1996)
Purbeck, Dorset	0.3	1882–1962	May and Heaps (1985)
White Nothe, Dorset	0.22	1882–1962	May (1971)
Barton-on-Sea, Hampshire	1.9	1950–80	Barton and Coles (1984)
Highcliffe, Hampshire	0.27	1931–1975	Univ. Strathclyde (1991)
Undercliff, Isle of Wight	0.05		Hutchinson (1991)
Blackgang, Isle of Wight	5		Clark et al. (1995)
Chale Cliff, Isle of Wight	0.41	1861–1980	Hutchinson et al. (1981)
Shanklin, Isle of Wight	0.68	1907–81	Clark et al. (1991)
Seven Sisters, Sussex	0.51	1873–1962	May (1971)

Location	Rate	Reference
Fairlight Glen, Sussex	1.43	Robinson and Williams (1984)
Beachy Head, Sussex	0.9	May and Heaps (1985)
Warden Point, Kent	1.5	Hutchinson (1973)
Studd Hill, Kent	1.5	So (1967)
Beltinge, Kent	0.83	Hutchinson (1970)
North Foreland, Kent	0.19	May (1971)
Walton-on-Naze, Essex	0.52	Hutchinson (1973)
Covehithe, Suffolk	5.1	Steers (1951)
Southwold, Suffolk	3.3	Steers (1951)
Pakefield, Suffolk	0.9	Steers (1951)
Dunwich, Suffolk	1.6	So (1967)
Runton, Norfolk	0.8	Cambers (1976)
Trimmingham, Norfolk	1.4	University of Strathclyde (1991)
Crome–Mundesley, Norfolk	4.2–5.7	Mathews (1934)
Marl Buff Kirby Hill, Norfolk	1.1	Hutchinson (1976)
Hornsea–Withernsea, Holderness	1.8	Pethick (1996a)
Withernsea–Kilnsea, Holderness	1.75	Valentin (1954)
Flamborough Head, Yorkshire	0.3	Mathews (1934)
Robin Hoods Bay, North Yorkshire	0.31	Agar (1960)
Saltwick Nab, North Yorkshire	0.04	Agar (1960)
Whitby (W), North Yorkshire	0.5	Clark and Guest (1991)
Whitby (E), North Yorkshire	0.19	Agar (1960)
Runswick Bay, North Yorkshire	0.27	Rozier and Reeves (1979)
Port Mulgrave, North Yorkshire	1.12	Agar (1960)
Crimdon–Blackhall, Durham	0.2–0.3	Rendel Geotechnics (1995)

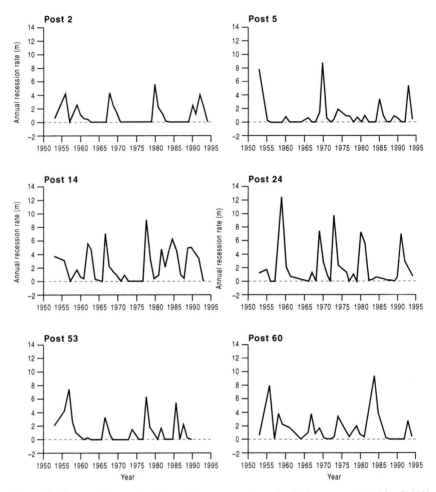

Figure 6.2   Representative annual cliff recession measurements on the Holderness coast, UK (after Pethick' 1996a).

tends to remove sand from the southern arm of the small coves formed by a single failure, so that a fresh failure occurs slightly south of the previous one. Failure coves tend to form sequentially southwards; thus the rate of recession at a particular point on the coast varies periodically depending on whether a headland or a bay is 'passing' the point of reference. The recession rate, therefore, varies from zero to over 10 m per year over a 5–8 year period.

Other clifflines are characterized by less frequent, episodic events. For example, the Isle of Wight Undercliff is a large pre-existing landslide complex that has been affected by various ground movement events. These events are probably the consequence of the slow degradation of the area and a very gradual decline in overall stability. The nature of ground movements within the Undercliff has been established by a search through historical documents, local newspapers from 1855 to 1994, local authority records and published scientific research (Lee and Moore, 1991; Moore et al., 1995). Over 200 events were identified that could be separated into two distinct groups: subsurface movements associated with the deep-seated creep of the entire landslide complex, surface or superficial movements arising from the erosion or failure of steep slopes and the differential movement and potential collapse (vents) between landslide blocks. Some parts of the Undercliff are very sensitive to rainfall events, while others appear only to show signs of movement during extremely rare conditions (Lee et al., 1998). The Undercliff also provides a good illustration of the concept of metastable equilibrium, with the very slow degradation process gradually reducing stability to a condition in which an abrupt change to a new form ultimately becomes inevitable; that is, the pre-existing landslide is replaced by a new sequence of first-time failures.

Although these and other clifflines have probably maintained a characteristic form, the rate of change has undoubtedly not remained constant over the last 1000 years (see table 2.1 for an interpretation of phases of instability on inland and coastal slopes). For example, the period from 1700 to 1850 has been associated with an increase in the reported incidence of major coastal landslides in eastern and southern England (Brunsden and Lee, in press), including the following:

1   the 1682 landslide at Runswick, North Yorkshire, when the whole village slipped into the sea (Young and Bird, 1822);
2   the 1737 landslide at The Spa in Scarborough's South Bay (Schofield, 1787);
3   the major failure in 1780, which destroyed the main road into Robin Hoods Bay, North Yorkshire and two rows of cottages;
4   the great landslide at the Haggerlythe, Whitby on Christmas Eve 1787, which resulted in the destruction of five houses and led to 196 families being made destitute (Anon. 1788);
5   the landslide of 1792 on the north-west of the Isle of Portland, which involved more than a mile of cliff and is believed to have been one of the largest coastal landslides to have occurred in historical times (Hutchins, 1803);
6   the major reactivation of parts of the Isle of Wight Undercliff, at Gore Cliff in 1799, and in The Landslip in 1810 and 1818 (see, e.g., Hutchinson, 1991);

7   the 1829 landslide at Kettleness, North Yorkshire, when the whole
     village slid into the sea, with the inhabitants having to be rescued
     by alum boats lying offshore (see e.g., Jones and Lee, 1994).
8   the great landslides on the North Norfolk coast near Overstrand of
     1825 and 1832 (Hutchinson, 1976);
9   the famous Bindon landslide, east of Lyme Regis, of Christmas Eve
     1839 (Conybeare et al., 1840) and the Whitlands landslide of 1840
     (Pitts, 1983).

Similar events have occurred since this period, such as the Holbeck
Hall landslide of 1993 in Scarborough (Lee, 1999), but they have been
much rarer. There is, however, no record of cliff recession rates during
the period 1700–1850, as it took place before the production of the
first Ordnance Survey maps. It can be argued that the period would
have been characterized by accelerated recession, but this is specu-
lation. Indeed, it is possible that the recession rates on some cliffs were
lower than today if, as is widely believed, beaches were more extensive
and larger, and could provide greater protection against wave attack at
the cliff foot.

The apparent stability of Chesil Beach, Dorset, suggests that this
famous shingle barrier has been in the consolidation phase. A comparison by Carr and Gleason (1972) of Sir John Coode's 1846–53 survey
of the beach (Coode, 1853) with longitudinal sections and cross-
sections made in 1967–9 suggested that there appears to have been no
clear changes to crest height or crest recession (table 6.2). However,
Carr and Seaward (1991) have noted a slight deterioration in the
correlation between crest height and longshore position over successive
surveys in 1852, 1969 and 1990.

The stable form and absence of overwashing behaviour, at least since
the early nineteenth century, suggests the beach has been in the
consolidation domain. However, there is some evidence that the beach
was more vulnerable to breaching and overwashing prior to that date:
'strong winds from the south east cause the sea to break through the
beach. Winds from the north west then restore the beach' (Leland,
1546); 'when the south wind rises (the beach) gives and commonly
cleaves asunder; but the north wind, on the contrary, binds and
consolidates it' (Camden 1590).

This might suggest that the beach has switched between consolidation and breakdown in the past. Indeed, Redman (1852), Codrington
(1870) and Groves (1875) all believed that the beach was higher when
they wrote than at earlier times, consistent with the data of Lilly (1715).
However, this evidence is by no means widely accepted as being a
reliable indicator of beach behaviour since the sixteenth century (Carr

Table 6.2 Measured trends in Chesil Beach from 1852 to 1990

Period	Trends	Sources
1852–1968/9	• Crest height increased by around 1 m between Abbotsbury and Wyke Regis • Crest height fell by 0.3 tm at Chiswell • Crest recession was less than the plotting error, except opposite Portland Harbour	Coode (1853) Carr and Gleason (1972)
1968/9–1977/8	• Crest height fairly constant between Abbotsbury and Langton Herring • Crest height increased by 0.7 m (up to 1.2 m) between Langton Herring and Tidmoor Point • Crest height variable towards Portland	Carr (1980)
1968/9–1990	• Crest height lower by, on average, around 0.5 m along the entire beach • Crest recession – no clear trend, as changes were considered to be primarily the result of erosion of the upper beach face • A line of concrete blocks at Abbotsbury appear to indicate 13 m of recession between 1948 and 1991, i.e. an average rate of 0.25 m yr^{-1}	Carr and Seaward (1990, 1991)

and Gleason, 1972; Carr and Blackley, 1974). Of particular note is the reported destructive nature of south-east winds; that is, from Portland harbour. It is possible that construction of the Portland breakwaters between 1850 and the late 1890s may have helped reduce the threat of breaching by waves from the harbour side.

Most estuaries have developed over the last 6000 years or so, after the postglacial rising seas flooded pre-existing deep river valleys or, as in Scotland, glacial troughs. These drowned valleys quickly began to infill with sediment carried by the rivers and from marine sources, developing the complex range of mudflats, saltmarshes, dunes, gravel beaches and subtidal channels that characterize British estuaries. Many estuaries are characterized by a long-term steady state balance between tidal processes and form and, hence, can be regarded as being in the consolidation domain. This balance can be expressed in terms of the following relationships (Pethick, 1994):

1 *Estuary length and tidal wavelength* (a function of mean depth). If mean depth increases, either as a result of sea-level rise, dredging or land reclamation (which removes intertidal areas and, hence, effectively increases mean depth) the estuary nodal point will migrate seaward, reducing velocities in the channel, allowing increased intertidal deposition and restoration of the stable form.
2 *Intertidal area and tidal prism.* As the intertidal area decreases, so there will be a decrease in tidal prism (O'Brien, 1931). This relationship may reflect the attainment of a critical current velocity for a stable estuary mouth of around 1 m s^{-1} (Pethick, 1994). Thus, the mouth will attain an area that will provide velocities and bed shear stress levels at, or immediately below, this critical erosion threshold (Escoffier, 1940; Goa, 1993).

The continual small-scale adjustment between process and form in estuaries was described by Wright et al. (1973): 'neither the channel morphology nor the tidal properties can be explained solely in terms of each other, though the two are mutually dependent. Simultaneous coadjustment of both process and form has yielded an equilibrium situation in which further adjustment is nonadventageous.' Evidence for the development of stable estuary forms was presented by Pethick (1994), based on a survey of 13 East Coast estuaries. Nine of these estuaries plotted within the 95% confidence limits of the theoretical relationships between length and depth (tidal wavelength) and intertidal area and tidal prism (figure 6.3). It follows that, over the last 1000 years, these estuaries, and probably many others around the coast, have been adjusting their form in response to changes in

(a)

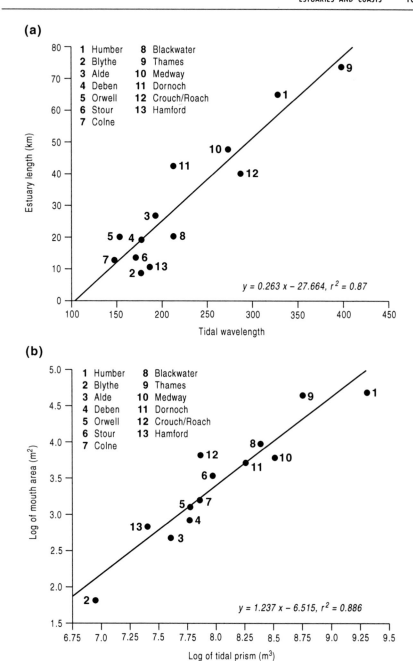

(b)

Figure 6.3   The morphology of East Coast estuaries (after Pethick, 1996b). (a) Tidal lengths approximate 0.25 tidal wavelength, suggesting resonant systems dependent upon shallow-water intertidal areas. (b) The relationship between the cross-sectional area at the mouth and the tidal prism.

depth, intertidal area and the cross-sectional area at the mouth. These adjustments have had the net effect of maintaining stable, steady state forms.

Within this pattern, many estuaries have probably oscillated between two forms characterized by alternating erosion and accretion of the intertidal areas. Initially, estuaries would have been relatively deep and wide, allowing the crest of the tidal wave to travel faster than the trough, which progresses more slowly in the shallower water. This leads to an asymmetry in the tidal wave, with a steep flood limb of short duration, while the ebb duration increases. Such *flood dominant* estuaries are often associated with a net input of sediment from the sea. The estuary becomes a sediment sink with rapid deposition rates on the intertidal areas, which gradually increase their elevation. In time, the channel cross-section defines an estuary in which the mean water depth is greater at low tide than at high tide (the high tide cross-section is dominated by shallow water over intertidal flats, while the low tide cross-section is confined to the deeper subtidal channel). In this condition, the asymmetry of the tidal wave is reversed, with the passage of the tidal wave crest slower than the trough (Dronkers, 1986). An *ebb dominant* estuary acts as a sediment source and is characterized by erosion of the intertidal areas, gradually returning the channel to a flood dominant form. Figure 6.4 provides an example of this type of oscillating behaviour, with the cross-section of the River Colne apparently having changed from an ebb dominant to a flood dominant form in the first half of this century (Pethick, 1996b).

A similar process may have occurred in the Humber where changes in sediment supply appear to have resulted in significant changes in estuary character. Pethick (1993) used navigation charts to assess the physical development of the Humber over the last 150 years. Since 1850 the estuary has been steadily decreasing in size, as mudflats and intertidal saltmarshes have formed and the channel has slowly filled with sediment. The average rate of accumulation has been 2 Mm^3 a year, or a vertical accretion of 6.5 mm a year. However, this disguises major changes in the pattern of sedimentation, with the estimated rates declining from 7 Mm^3 a year between 1900 and 1920 to 1.5 Mm^3 by 1970, a decrease of 78%. Analysis of the most recent charts suggests that the level of accumulation has fallen, so that in the 1980s zero net accumulation occurred.

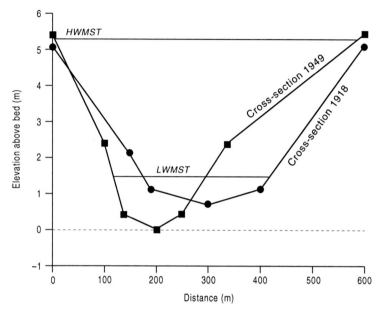

Figure 6.4 Channel change in the Colne estuary 1918–1949, suggesting development from flood dominant to ebb dominant forms (after Pethick, 1996b).

6.2.3 Adjustment

The East Cliff–Black Ven area, east of Lyme Regis, provides a clear example of how a coastal system has adjusted to changing controls within the last 1000 years. The area is a remnant of a once more extensive periglacial slope that formed between Lyme Regis and Charmouth (the Black Ven–Spittles–East Cliff landslide complex). These relict slopes were covered with relict landslide debris and head deposits, formed during past phases of slope instability.

Renewal of landslide activity has occurred in response to the gradual onset of marine erosion at the cliff foot. This is believed to have been relatively recent (within the last 1000–2000 years) and has led to the dramatic transformation of the Black Ven area and, in recent years, the Spittles from a degraded landslide cliff to one of the most actively unstable coastal cliffs in Britain. Marine erosion at the cliff foot removed the periglacial debris slopes, created the near-vertical seacliffs and led to the progressive reactivation of the relict landslide systems. As marine erosion continued, the cliff profile steepened, making the system increasingly susceptible to mudsliding. In time, steady state mudslide activity became established under current environmental conditions (Chandler and Brunsden, 1995).

The western section of Black Ven probably formed during the Little Ice Age, around 1600–1700, while the Spittles only became a fully reactivated landslide system in 1986 (Brunsden and Chandler, 1996). The area of active instability probably spread rapidly because the materials along the pre-existing shear surfaces are at residual strength and, hence, readily destabilized by the toe unloading. In time, the entire pre-existing landslide system will be exhumed, facilitating the initiation of first-time failures and the development of a new cliff form.

Estuaries may also undergo significant adjustments that probably mark a shift from the consolidation domain. In figure 6.3, 4 of the 13 East Coast estuaries show pronounced changes from the theoretical steady state form. Pethick (1996b) records that the outer Blackwater channel, Essex, widened by over 1000 m between 1820 and 1992, and the intertidal area flattened and lowered. Over the period 1973–88, 200 ha of saltmarsh was lost, representing 23% of the total saltmarsh area (Burd, 1989). Pethick (1994) explained these adjustments as being a response to relative sea-level rise and the resulting increase in tidal prism. In this case, the gradual return to the steady state form of the consolidation domain appears to have been constrained by flood embankments, which prevent the onshore migration of the intertidal profile, so that the upper intertidal area is truncated as sea-level rise occurs.

In the Medway estuary in Kent, the removal of saltmarsh sediment for use in brick-making in the eighteenth, nineteenth and early twentieth centuries probably triggered a period of significant erosion. Pethick (1994) suggests that this led to increased tidal velocities and wave generation within the estuary. Marsh loss accelerated during the early twentieth century, partly due to the effects of storm tides in 1897 and 1901 that caused numerous seawall breaches (Pye and French, 1993). Around 1500 ha of saltmarsh has been lost (Kirby, 1990; IECS, 1993). Pethick (1994) speculated that the estuary might recover from these changes (i.e. switch to the steady state condition of the consolidation domain) as intertidal deposition reduces the tidal prism, although the reduced inlet size might mean that artificial recharge will be needed to initiate the process.

By 1840, land reclamation in the Blyth estuary, Sussex, had reduced the former intertidal extent by 1100 ha and formed a narrow parallel-sided channel (Pethick, 1994). Collapse of the inner embankments in the 1940s resulted in the re-creation of 250 ha of intertidal land (Beardall et al., 1991). However, the high tidal discharges and velocities, along with waves generated across the large expanse of water within the inlet, have prevented saltmarsh development and a return to the steady state form.

There have also been marked changes to the Walton Backwaters, Essex between 1847 and 1995, with a shallowing of the Pye Channel by around 4 m and a 17% reduction in the intertidal area (Rampling et al., 1996). These changes may reflect the oscillation from flood to ebb dominance described earlier, in response to a reduction in the tidal prism due to land reclamation. However, over the same period, Stone Point at the channel mouth has eroded southwards at 3 m yr^{-1}, widening the inlet. Inlet narrowing would normally be the expected response to a reducing tidal prism. The observed widening suggests that there may have been a significant change in one of the main controls on the estuary form. Indeed, Pethick (1995) suggests that has been a regional decline in sediment availability such that the estuary has not been able to maintain its equilibrium form under conditions of rising relative sea level.

Perhaps the best known examples of estuary changes have occurred in the River Dee, where there has been a well documented history of land reclamation (around 5000 ha since 1732; Fahy et al., 1993), channel change, abandoned ports, localized erosion and extensive growth of new saltmarsh on the upper intertidal surfaces (see, e.g., Marker, 1967; Pye, 1996). For example, Parkgate had been a bathing resort, but by the late 1930s the sandy beach had started to be colonized by saltmarsh vegetation. By 1947 the entire beach was colonized, with the beach covered by a 450 m wide saltmarsh. This marsh had increased to 1200 m wide by 1967. Marker (1967) records accretion rates of 58 mm yr^{-1} between 1938 and 1947, and 15 mm yr^{-1} between 1947 and 1963. It is possible that the accelerated accretion in the twentieth century was caused by a reduction in tidal prism and the ebb tide current velocity following canalization, embanking and reclamation (Pye, 1996). Channel training may have also encouraged the progressive deposition by controlling saltmarsh erosion.

Over the last 100 years or so, the colonization by *Spartina anglica*, a hybrid grass, has transformed the intertidal areas of many estuaries. It first appeared in Southampton Water in 1870 but spread rapidly (it was often planted for shore stabilization and land reclamation), reaching Poole Harbour in 1899 and the Solway Firth by 1958 (Davidson et al., 1991). It spread across mudflats, tapping sediment and led to the growth of extensive saltmarsh systems. In recent years there has been a decline in *Spartina* marsh ('die-back'), although the reasons remain unclear. In the Wash, for example, there was a decline from 2000 ha in 1958 to 200 ha by 1973 (Doody, 1984).

The most dramatic changes along the coastline have occurred at

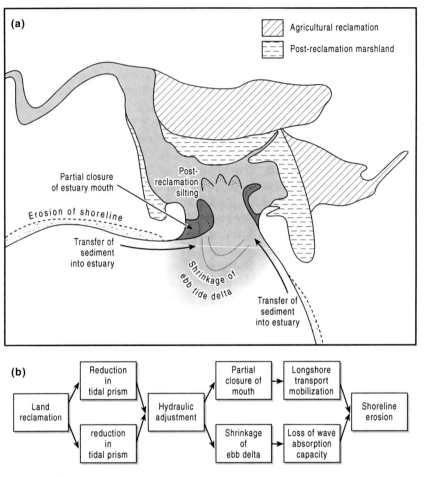

Figure 6.5 A framework for explaining morphological changes within estuaries. (a) A cartoon of human-induced changes. (b) A sequence of events from land reclamation to shoreline erosion (after Carter, 1992).

estuary mouths, where changes in tidal prism (often associated with land reclamation) have led to shrinkage of ebb tide deltas and partial closure of the inlet by features such as spits, bars and dune systems (figure 6.5). The following examples serve to illustrate the types of changes that have been characteristic of many estuary mouth areas:

1   *Bridgewater Bay.* Kidson (1963) describes the growth of a series of spits into the bay, following a major breach of the coastline in 1783. This breach had cut off a large area of land at the mouth of the River Parrett, creating Stert Island. By the 1850s a spit had

begun to grow from Stert Point on the mainland, and by 1902 it had enclosed Fenning Island, an extension of around 1000 m in 50 years.

2 *Poole Bay.* The rapid growth of the Studland dune complex, probably since 1600, is reported by Robinson (1955) and Goudie and Brunsden (1997). The coast has built seawards by around 850 m in the north and 350 m in the south as a series of dune ridges and intervening depression (slacks). The westernmost ridge, Third Ridge, was probably completed by 1720. The second, Second Ridge, was fully developed by the 1840s, while First Ridge was formed by 1900. A further ridge, Zero Ridge, has been developing since the 1950s.

3 *Christchurch Harbour.* The northeasterly 1 km extension of a spit across the harbour entrance between the 1840s and 1900s forced the Avon and Stour river channel (the Run) close inshore, where it ran parallel with a northern spit (Robinson, 1955). A breach occurred to the southerly spit in 1924, through which the river cut a new channel. Subsequent spit growth had forced the opening around 400 m northeastwards by 1926. A further breach occurred in 1935.

4 *Beaulieu Mouth.* Since around 1907, Warren Farm Spit has grown from Gravelly Marsh, on the west side of the river mouth, to Needs Oar Point, a total length of around 1400 m (Hooke and Riley, 1991). Over the same period a saltmarsh (Gull Island) developed to the east of Needs Oar Point. Hooke and Riley (1991) report that the Beaulieu estate believed that much of the change occurred in the winter storms of 1953–4 and coincided with commencement of offshore aggregate dredging.

5 *Chichester Harbour.* There have been significant changes to East Head, the spit on the east side of the harbour entrance, over the last 100 years or so (Searle, 1975; Webber, 1979). Prior to the 1870s the spit ran on an east–west line, a continuation of the open coast orientation. Between 1875 and 1887 East Head began to swing northwards to its current orientation. These changes probably reflect a variety of long-term and man-induced changes in Chichester Harbour and on the open coast, including the decline in sediment supply and longshore transport due to coast protection works between Selsey Bill and East Head. Webber (1979) estimated that the sediment supply to East Head from the open coastline declined from around 70 000 m^3 in the mid-nineteenth century to 6500 m^3 by 1979. This decline was caused by the progressive protection of the open coastline from around 1846 onwards.

6   *Pagham Harbour.* In 1587 the harbour entrance had been flanked by two spits. This had changed to a single shingle barrier almost completely across the mouth by 1672 (Robinson, 1955). The distal end of this barrier extended eastwards by over 500 m between 1785 and 1843, forcing the harbour entrance channel to cut a low clay cliff along its northern side. This cliffline retreated over 150 m during this period, causing the loss of a mill. The barrier extended a further 400 m between 1843 and 1874, during which time the barrier also rolled landwards by over 100 m across the harbour mudflats. Since the 1870s, the barrier has been managed to control flooding and erosion risks.

7   *Orford Ness.* The periodic southerly extension and contraction of Orford Spit is described by Carr (1969). Over the period 1812–21, there was 2900 m fluctuation in the total length of the spit, with the most southerly position of the distal end occurring in the early nineteenth century (Cobb, 1957). Estimated rates of growth include 183 m yr^{-1} (1804–12), 69 m yr^{-1} (1867–80) and 64 m yr^{-1} (1962–7).

8   *Spurn Head.* The historical development of Spurn Head has been described by de Boer (1963, 1964) as involving a 250 year cycle of growth and breaching of the neck, with each new spit growing to the north-west of its predecessor. However, IECS (1994a) have proposed a different interpretation to explain the historical changes, suggesting that the spit has transformed from a shingle to a sand-dominated feature over the last three centuries because of the removal of shingle for ship ballast.

9   *Aln Estuary, Northumberland.* Collins (1993) describes the development of dune systems and saltmarshes at the mouth of the Aln between the early seventeenth and late nineteenth centuries. In 1806 the Aln changed channel, cutting through a low sand and gravel ridge, leaving a small island, Church Hill, in mid-channel. In the mid-nineteenth century a sandy spit and dune complex extended northwards from the south shore of the estuary mouth to envelop Church Hill, closing off the former channel.

10 *Harlech, Wales.* It is believed that a major sand dune and saltmarsh complex developed at the mouth of the River Dwyryd since the construction of Harlech Castle in 1286. This castle has a watergate and is known to have been a port in the thirteenth and fourteenth centuries (Steers, 1946).

6.2.4 Breakdown

The decline in sediment availability since 5000–6000 BP is believed to have had a marked impact on many coastal systems, especially on the shoreline orientation and planform of shingle beaches. At a broad scale, scarcity of sediment can lead to cell fragmentation, the breakdown of established sediment cells into smaller units. On the West Dorset coast, for example, Lee and Brunsden (1997) demonstrated how this process had resulted in the development of a series of pocket beaches separated by rocky headlands, probably within the last 1000 years or so. Historical chart and photograph evidence suggests that the four discrete shingle transport cells (cell 1, Lyme Regis to Golden Cap; cell 2, Golden Cap to Doghouse Hill; cell 3, Doghouse Hill to the West Bay Piers; cell 4, the West Bay Piers to the Isle of Portland) are the remnants of what once may have been a continuous shingle beach extending from Lyme Regis to the Isle of Portland. Prior to the construction of The Cobb at Lyme Regis (it was connected to the foreshore in 1754) this beach may have even extended westwards into East Devon (Brunsden, 1991). However, the long-term effects of differential erosion of the more resistant headlands (Golden Cap, Doghouse Hill to Thorncombe Beacon) and the softer rocks in the intervening bays has gradually created the present-day sequence of pocket beaches, which are only episodically linked by shingle transfers. Possibly the strongest evidence for this fragmentation is the relict beach deposits beneath Doghouse Hill (Bray, 1996). These comprise substantial quantities of large (20–80 mm diameter) well-rounded chert and flint pebbles that could only represent a former, now depleted, beach and shingle transport pathway.

The fragmentation of the continuous beach into discrete pocket beaches is a relatively recent phenomenon. Indeed, it is likely that the coastline has changed from partial fragmentation to almost complete fragmentation within the last 200 years or so. The likely sequence of events is as follows (figure 6.6):

1   *Pre-1740.* There may have been an almost continuous sand and shingle beach extending from East Devon to the Isle of Portland. The shingle beach would have been supplied by the eroding cliffs of East Devon, between Sidmouth and Charmouth, with the dominant westerly winds resulting in a net drift towards Chesil Beach. At this time, The Cobb was detached from the shoreline and would not have significantly affected the littoral drift. Around this time, William Jessop (1805) wrote:

(a)

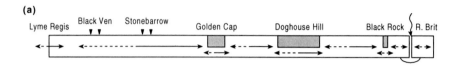

(b)

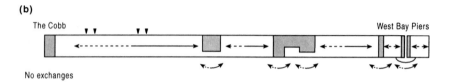

(c)

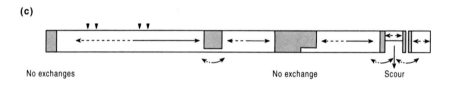

(d)

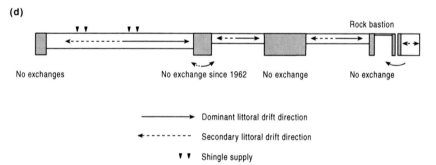

Figure 6.6 A schematic model of the fragmentation of coastal cells on the West Dorset coast (modified from Lee and Brunsden, 1997). (a) Pre-1740; (b) 1756 – connection of The Cobb to the Foreshore (1756) and construction of West Bay Piers (1744); (c) 1850 – closure of the Doghouse Hill–Thorncombe Beacon cell boundary; (d) 1990 – construction of bastion at Black Rock.

the causeway which connects it [The Cobb] with the Land is only one or two feet above high water. The Shingle which is constantly travelling from West to East along the Coast is washed over this Causeway and except what is retained by the Jetties which are projected to preserve the Town, moves on until an immense Collection of it is accumulated at the Chesil Bank which connects the Isle of Portland with the mainland.

2   *1756*. The construction of the West Bay piers in 1744 and the connection of The Cobb to the shoreline in 1756 both restricted the free movement of shingle along the foreshore. This was a very significant time for the evolution of Chesil Beach, as it became at least partially cut off from its major contemporary sources of shingle in East Devon and east of Lyme Regis (Black Ven and Stonebarrow). Towards the end of the eighteenth century, headlands and rock ledges below Golden Cap and Thorncombe Beacon had begun further to restrict shingle transport along the foreshore, increasing the fragmentation of the coastal cells.

3   *Around 1850*. By this time, the Doghouse Hill–Thorncombe Beacon headland would probably have formed a permanent barrier to the eastwards drift of shingle supplied from the eroding cliffs of Black Ven and Stonebarrow. In addition, the West Bay piers would have reduced the potential for shingle exchanges between Chesil Beach and the pocket beaches to the west.

4   *Late twentieth century*. The occurrence of landslide boulder arcs below Golden Cap created a temporary barrier to shingle movement between Charmouth and Seatown beaches. Closure of this cell boundary has taken place in 1914, 1949 and, most recently, in 1962. The construction of a rock bastion on Black Rock, West Bay in the early 1980s has prevented significant shingle exchanges between Eype Beach and West Beach.

It is believed that there is no significant shingle exchange between adjacent cells at present. Mudslide lobes below the Golden Cap headland have prevented shingle transport since around 1962 (and could continue to do so indefinitely; Bray, 1996). Limestone and sandstone boulders between Doghouse Hill and Thorncombe Beacon have probably prevented exchanges since between 1787 and 1850, and possibly earlier. The construction of piers across the shingle beach at West Bay to allow access into the harbour and control the course of the River Brit has created an artificial barrier between West Beach (cell 3) and East Beach (cell 4).

On the East Devon coast, cell fragmentation has occurred over a similar timescale as a result of major landslides forming barriers to longshore shingle transport (Lee and Brunsden, 1999). The suggested sequence of events is as follows (figure 6.7):

1   *c.1800*. The connection of The Cobb to the shoreline in 1754 restricted the free movement of shingle along the foreshore, promot-

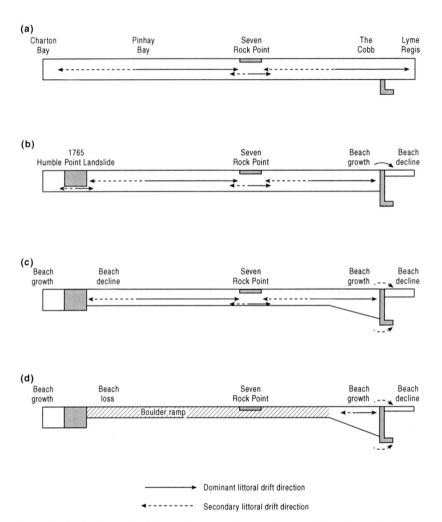

Figure 6.7 A schematic model of the development of the East Devon coast since the seventeenth century (modified from Lee and Brunsden, 1999).

ing the build-up of a large shingle and sand beach on the western side (Monmouth Beach). The 1765 landslide at Humble Point probably prevented, or significantly reduced, the longshore transport of shingle from Charton Bay into Pinhay Bay which, at this time, contained an extensive shingle beach. By 1787 the headland at Seven Rock Point may have become at least a partial barrier to longshore shingle transport from Pinhay Bay to Monmouth Beach.

2   *c.1850.* The 1840 Humble Point landslide (one of the largest landslide events on the British coast over the last 200 years) reinforced the sediment barrier between Charton Bay and Pinhay Bay, promoting the build-up of the present-day shingle berm behind the landslide debris. By around 1857, the bridged causeway to The Cobb had been replaced by a solid connection. It is likely that this work further reduced the already diminished sediment exchanges between Monmouth Beach and Town Beach, Lyme Regis.

3   *1900–1990s.* By the turn of the century, the shingle beach in Pinhay Bay appears to have been lost, replaced by the present-day boulder beach. Monmouth Beach has continued to build up behind The Cobb. Additional works on the causeway in the 1930s (sluices were installed and later concreted over) and 1959 (when the parapet was raised) ensured that shingle could not move from Monmouth Beach except in the most severe conditions.

As a result of these adjustments over the last 250 years or so, there have been marked changes in the character of the shingle beaches along this coast. Charton Beach has grown, as the shingle output from the bay has declined to zero because of the barrier caused by the Humble Point landslide debris. Pinhay Beach has disappeared, as the shingle input from Charton Bay declined to zero while the output alongshore to Monmouth Beach continued. Monmouth Beach has built up behind The Cobb, as shingle outputs to Town Beach have declined to almost zero while still receiving an input from Pinhay Bay.

At a more detailed level, scarcity of sediment leads to cannibalization; that is, the reworking of sediment within a beach or barrier structure, leading to a new morphology and behaviour. Where the down-drift sediment supply is sufficient to fulfil the available longshore power, drift-aligned structures will be developed and maintained (Davies, 1972), associated with longshore extension of the beach. However, where sediment supply is insufficient, the sediments already within the beach system are liable to erosion and transport alongshore. In time, this internal redistribution of sediment may create a swash-aligned beach in which the breaking wave approach is near-normal at all points (no potential transport), or where sediment grading develops so that actual transport is zero at all points (Carter, 1988); that is, the long-term condition is no net longshore movement. For example, there have been a number of significant changes to Monmouth Beach, Lyme Regis over the last 200 years or so (Lee and Brunsden, 1999). These include the following:

- *Beach widening.* The beach has grown outwards at The Cobb, since the connection of The Cobb to the shoreline in 1754 restricted the free movement of shingle along the foreshore. There has been a seaward advance of HWM adjacent to The Cobb of around 150 m over the last 220 years, indicative of a substantial growth of the beach in this area. Posford Duvivier (1990) estimated that the beach had accumulated at rates of up to 2900 m^3 yr^{-1}.
- *Beach shortening.* The western end of the beach has migrated towards The Cobb by over 600 m since 1854 and, perhaps by as much as 2500 m since 1787. Most of this migration occurred before 1946, after which time the western end of the beach appears to have fluctuated backwards and forwards by 90 m.
- *Beach steepening.* As HWM at The Cobb has advanced seawards, LWM has retreated by around 50 m. These changes have resulted in an increase in the beach face slope angle from 2° in 1854 to 6° in 1997, suggesting a significant coarsening in the size of the beach face material.
- *Beach orientation.* Although the eastern end of the beach appears to have built out parallel to the former shoreline, there has been a notable reorientation of the western end by around 10 degrees.

Figure 6.8 presents a simple model that explains the changes at Monmouth Beach in the context of limited sediment supply to the beach (because of the landslide barriers further east on the Devon coast, described earlier). As the beach has built up against The Cobb, it has shortened, that is, the beach accumulation in the east has been at the expense of the western end of the beach. It is likely that this cannibalization was concentrated between the 1850s and the 1940s, as indicated by the rapid shortening of the western end of the beach during this period.

In their work on shingle barriers, Carter and Orford (1993) and Orford et al. (1996) have suggested that the transition between barriers in the consolidation and breakdown domains was expressed as a change from crestal overtopping to crestal overwashing. Thus, the failure of the shingle barrier at Porlock, Somerset during a storm in October 1996 has been interpreted as indicating the switch to breakdown (Jennings et al., 1998). Jennings and Smyth (1990) have suggested that the shingle beaches of the Sussex coast are predominantly relict features, inherited from past phases of rapid sea-level rise after the end of the last glaciation, between 10 000 and 5 000 years ago. Initially, beach material would have been moved onshore from seabed sources, under the influence of rising sea level (the initiation and consolidation domains with an offshore–onshore dominant sediment transport

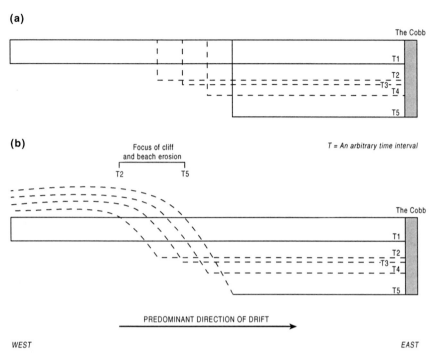

Figure 6.8   A model explaining contemporary changes at Monmouth Beach, Lyme Regis (after Lee and Brunsden, 1999).

regime). Once all the available seabed shingle had been transferred onshore, a longshore-dominant sediment transport regime was established (the adjustment domain). In time, this would have resulted in the gradual depletion of beaches as shingle was moved progressively down-drift (eastwards towards structures such as the Crumbles at Eastbourne and Dungeness) at a faster rate than it was supplied by cliff recession (breakdown). At the Crumbles, three periods of growth have been suggested, at around 9000 BP, 3800 BP and, most recently, during the twelfth century (Jennings and Smyth, 1990), suggesting that the transition between beach growth and breakdown at the Crumbles occurred within the last 300 years.

6.2.5   Abandonment

Cases of 'natural' abandonment of landforms during the last millennium are restricted to a number of settings where estuary mouth dune/saltmarsh complexes have formed in front of previously eroding clifflines (e.g. at Harlech and in the Aln estuary, as described earlier) or where relative sea-level fall has stranded features such as salt-marshes, as in some Scottish estuaries. However, a more widespread trend has been 'artificial' abandonment as a direct result of coastal defence operations. Erosion control structures have been built at the foot of many clifflines. However, unless these works were accompanied by effective slope stabilization measures they have continued to degrade towards long-term angles of stability. Whilst slope degradation behind defences generally involves relatively small and minor events, large-scale dramatic events do occur and can result in considerable loss of land. The 1993 Holbeck Hall landslide, Scarborough, for example, led to the destruction of the hotel and sea walls below, with a loss of around a 95 m width of land (Clark and Guest, 1994; Lee, 1999). At Overstrand, Norfolk around 100 m of clifftop land was lost during a three year period between 1990 and 1993. The slope toe had been protected by wooden breastwork defences (Frew and Guest, 1997).

6.3   The Impact of Humans

It would be difficult to over-state the effect that man has had on coastal and estuarine landforms, especially over the last 250 years or so. Man's impact has been a factor in many of the changes described in this chapter. In estuaries, land reclamation and dredging operations have led to changes in tidal prisms resulting in morphological adjustment, especially around the mouth. Hydraulic adjustments in the estuary will have also caused the shrinkage of the ebb tide delta. These delta features often comprise shallow bars that extend in front of the adjacent coast. Shrinkage of the delta can result in loss of wave absorption capacity on the open coast and, hence, accelerated erosion.

Collapse of an ebb tide delta may have been a factor in the onset of erosion at the former port of Dunwich, Suffolk (Pethick, pers. comm.). It is possible that reclamation in the Blyth and Dunwich estuary triggered the southwards growth of a shingle spit across the mouth and the collapse of the tidal delta. The accelerated erosion and loss of much of Dunwich village and its churches is well documented (Chant, 1986; Bacon and Bacon, 1988). For example, Gardner (1754) records that by 1328 the port was virtually useless, and that 400 houses, together with windmills, churches, shops and many other buildings, were lost in one night in 1347.

Construction of flood embankments has constrained the 'natural' tendency of an estuary to establish a steady state balance between morphology and process. As a consequence, widespread saltmarsh erosion has occurred in front of these defences. More than 20% of the saltmarsh area in Suffolk, Essex and north Kent was lost between 1973 and 1988 (Burd, 1992). In addition, flood embankments prevent the saltmarsh/mudflat profile from 'rolling' onshore in response to relative sea-level rise. As a result, many intertidal profiles have reduced in width, causing significant loss of habitat (i.e. 'coastal squeeze'; English Nature, 1992). Keyhaven and Pennington marshes in the Solent, for example, have narrowed by 500 m and 225 m, respectively, over the last century (Hooke and Riley, 1991). Elsewhere, as in the Dee estuary, land reclamation and channel training has resulted in accelerated accretion. The planting of *Spartina* to provide shoreline protection or assist reclamation has also promoted saltmarsh growth, at the expense of other intertidal landforms and habitats.

On the open coast, human intervention may lead to significant changes in behaviour and stability (Bray and Hooke, 1995). Indeed, the effects of coastal engineering works, for example, may be compounded over time and eventually threaten the integrity of many of the dynamic landforms in a cell (figure 6.9). Coast protection works can interfere with and stop sediment supply from cliffs. Over the last 100 years or so, some 860 km of eroding coast in England (predominantly soft cliffs) have been protected (MAFF, 1994). This probably represents 75% of the soft cliff resource and is likely to have had a dramatic effect on the sediment budgets of many coastal cells. Pye and French (1992) estimated that only 250 km remain unprotected. The resulting beach depletion downdrift may lead to the need for further coastal defence works, again with impacts on the sediment transport system operating in the cell. In time, transport pathways become broken, landforms functioning as net stores may begin to erode and become finite open systems, and sinks become finite closed systems.

Dredging of sand and gravel has had significant impacts on some coastlines, most notoriously at Hallsands, Devon (Mottershead, 1997). In 1890 there had been an extensive shingle beach in front of the village, part of the Start Bay barrier complexes. However, between 1897 and 1900 around 400 000 m^3 of shingle was removed from the seaward side of the beach by a contractor, Sir John Jackson, for construction in the Royal Navy dockyards at Devonport. The beach levels declined rapidly, falling by 3–4 m by 1903, and wave attack led to the loss of several houses. A sea wall was built in 1904, although this was severely damaged by storms in 1917. Today, the once busy fishing village is deserted.

Significant coastal adjustments may also occur in response to the dumping of material on the foreshore. For example, on the Fife coast

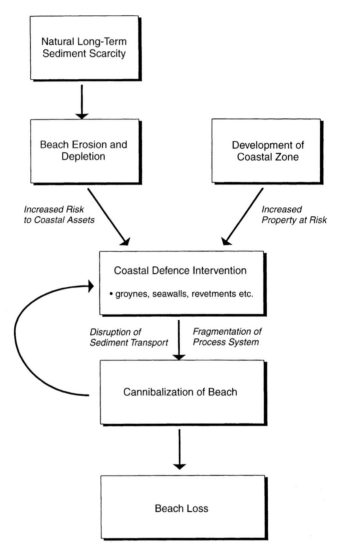

Figure 6.9 A summary of the 'knock-on' effects of coast protection works on shingle beaches (modified from Bray and Hooke, 1995).

excess mine waste from the Wellesley Colliery (mining began in the 1890s) was deposited on the shoreline (Saiu and McManus, 1998). By the 1950s the waste had been redistributed along the shoreline, forming an apron of material some 100 m wide in front of the colliery and causing siltation in Buckhaven harbour. Waste from the nearby Michael and Frances Collieries created headlands that extended 150 m out to sea. Since closure of the mines in the 1960s, the unprotected waste aprons have begun to erode, with up to 30 m recession recorded in a six month period. A similar pattern has occurred on the Durham coast, where the beaches were completely altered by many millions of tonnes of waste from Easington Colliery (Simm et al., 1996). The 'Turning the Tide' project is ambitious programme of coastal and beach restoration that is now under way in County Durham.

Coastal structures have often led to disruption or breakdown of longshore sediment transport pathways. For example, construction of the West Bay piers in 1744 and their infilling by Francis Giles in 1820 has effectively cut Chesil Beach off from a source of further shingle (Lee and Brunsden, 1997). Over a similar period, beach mining operations have probably reduced the shingle volume by between 1.5% and 5%. Despite these impacts, there appear to have been no significant morphological changes to the main section of Chesil Beach from Abbotsbury to the Isle of Portland (Carr and Seaward, 1991). However, the deterioration in correlation between crest height and longshore position since 1852 suggests that the beach may be showing signs of very slow breakdown. Lee and Brunsden (1997) suggest that the long-term future of the western sections of Chesil Beach, west of Abbotsbury, is one of gradual breakdown into pocket beaches and headlands. However, the enormous beach volume probably provides a buffer against these changes on this section and may ensure that breakdown will be a very slow process.

Coastal structures often cause a localized change in the coastline orientation and a subsequent reduction in longshore shingle transport potential (the rate of longshore transport is sensitive to coastline orientation). Both effects will gradually lead to a decline in the sediment supply to the downdrift beaches. This trend has been well documented on the Holderness coast. Using both historical map sources (1852–1952) and erosion post data (1953 to present day), previous research has identified that coastal defence works at Hornsea and Withernsea, built early this century and subsequently extended and improved, have had an effect on the rate of cliff recession, causing a decrease to the north and an increase to the south of the groyne fields (Valentin, 1954; IECS, 1994b; Posford Duvivier, 1993). The modification to the recession rate appears to have been greatest around Hornsea, where differences of between 1.9 and 2.4 m yr^{-1} have been recorded between the cliff at similar distances to the north and south of the defences (table 6.3; Posford Duvivier, 1993). IECS (1994b) analysed historical records and erosion post data, and concluded that the Hornsea and Withernsea defences had led to the development of a

Table 6.3 Average cliff recession rates north and south of existing protection works at Hornsea and Withernsea, Holderness (after Posford Duvivier, 1993)

Length (km)	*Hornsea*			*Withernsea*		
	North $(m\ yr^{-1})$	*South* $(m\ yr^{-1})$	*Difference* $(m\ yr^{-1})$	*North* $(m\ yr^{-1})$	*South* $(m\ yr^{-1})$	*Difference* $(m\ yr^{-1})$
2	0.38	2.81	2.43	0.26	0.97	0.71
3	0.60	2.65	2.05	0.45	0.85	0.40
4	0.66	2.59	1.93	0.65	1.14	0.40
5	0.65	2.51	1.86	0.79	1.23	0.40

zone of accelerated erosion which extended at least 10 km south of the groyne fields.

These trends are believed to reflect the development of a bay between the protected towns of Hornsea and Withernsea. In between these points erosion would continue, but would result in a bay in which sand would accumulate, and eventually this would act to stabilize the cliff, so that further erosion along the entire coastline would be halted. The IECS (1994b) analysis indicates that the rate of bay development south of Hornsea has been increasing over the last 40 years, with average annual recession rates of 1 m south of the groyne field in 1955 to nearly 3 m yr⁻¹ in the 1990s. Pethick (1996a) has suggested that the impact of these schemes on the shoreline orientation has been a critical factor:

> The change in shoreline orientation to the north of each defence has resulted in a decrease in the potential transport rate and consequently an accumulation of sand on the beach. This has led to the progressive masking of the free face of the cliffs by debris in both locations and cliff recession has now halted immediately to the north of the defences. To the south of each defence, the change in orientation of the shoreline also results in a reduction in the potential sediment transport rate, but here the input of sand is reduced due to the accumulation of sediment to the north. The result has been a net sediment deficit, a decrease in stored beach sediment and an increase in basal erosion. The increase in recession rates to the south of each defence has resulted in the formation of a marked asymmetric embayment which is progressively outflanking the defences as it migrates northwards.

Finally, it should be noted that coastal defence works have also had an impact on habitats. The significance of these losses was highlighted

Table 6.4 An estimated habitat loss/gain account for SAC/SPA and Ramsar sites in England and Wales (after Lee, 1998)

Habitat	Estimated loss (ha)	Estimated gain (ha)	Balance (ha)
Mudflat/sandflat	11 459	12 991	1 532
Saltmarsh	6 996	7 685	689
Shingle bank	238	110	−128
Sand dune	504	381	−123
Cliff top	133	0	−133
Soft cliff	0	0	0
Hard cliff	0	0	0
Wet grassland	3 214	0	−3 214
Reed bed	172	0	−172
Coastal lagoon	530	30	−500
TOTAL	23 246	21 197	−2 048

Predictions are based on a single 'best-guess' coastal defence scenario for the next 50 years, identified from a review of available Shoreline Management Plans (SMPs) and through a series of regional workshops.

in a review of possible future losses of internationally important conservation sites around the coast and England and Wales (Lee, 1998). Loss/gain accounts were developed for individual coastal cells in England and Wales, based on a 'best-guess' coastal defence scenario for the next 50 years (i.e. do nothing, hold the line, advance the line or managed retreat). If the 'best-guess' coastal defence policies are implemented and if the predicted coastal changes occur, a range of important habitat changes were considered possible (table 6.4). There could be a net loss of freshwater and brackish habitat of around 4000 ha, primarily wet grassland (c. 3200 ha) but also including significant areas of coastal lagoon (c. 500 ha) and reed bed (c. 200 ha). In addition, a net gain of intertidal (saltmarsh and mudflat/sandflat) habitats of around 2200 ha is predicted, with the gains associated with managed retreat (c. 12 500 ha) balancing the expected losses due to coastal squeeze and erosion on the unprotected coast. It is estimated that around 120 ha of sand dunes and around 130 ha of shingle bank habitats could be lost over the next 50 years.

6.4 The Nature of Future Changes

Throughout the last millennium, coastal landforms have adjusted to changes in the dominant controls – either 'natural' or man-induced. The observed or recorded changes can be explained in terms of the ongoing adjustment of persistent and apparently stable forms within a particular process domain to variations in energy inputs or sudden and dramatic changes as the landform passes into a different domain (e.g. the switch from the consolidation to breakdown domain reported for some shingle beach systems). Although five domains have been recognized, only three are of particular significance for the British coast in the last 1000 years – consolidation, adjustment and breakdown. This probably reflects a continuation of trends inherited from previous millennia. However, in future more systems are likely to move towards the spatial instability of the breakdown domain, primarily because of the continued decline in sediment availability and the effects of relative sea-level rise.

Bruun (1962) predicted that, in response to sea-level rise, shore profiles would respond by acting to maintain their morphology relative to still water levels. This would be achieved by translation of the landform landwards and upwards, with erosion at the landward end of the profile supplying material to raise the lower portion of the profile (assuming a net sediment balance). This basic idea appears to apply to all coastal landforms. The concept of landform migration provides an approach to interpreting the degree of future adjustment to sea-level rise. A range of responses are envisaged, depending on the local circumstances and interaction with coastal defence infrastructure:

1   No significant change to current behaviour; for example, a slight increase in hard rock cliff recession rates.
2   A constrained response, where landform migration is prevented by a defence line or topography; for example, a beach–sea wall system.
3   A partially constrained response, where landform migration is limited by a defence line or topography; for example, a partly defended estuary.
4   A free response, where landward, seaward or longshore migration can occur without constraint.
5   A major change to current behaviour; for example, breakdown of a shingle beach system, reactivation and subsequent transformation of a pre-existing landslide complex.

REFERENCES

Agar, R. 1960: Postglacial erosion of the north Yorkshire coast from Tees estuary to Ravenscar. *Proceedings of the Yorkshire Geological Society*, 32, 408–25.

Anon. 1788: *The Gazetteer and New Daily Advertiser*, 5 January 1788 (reproduced in *The Whitby Times*, 28 December 1894).

Bacon, J. and Bacon, S. 1988: *Dunwich Suffolk*. Colchester: Segment Publications.

Barton, M. E. and Coles, B. J. 1984: The characteristics and rates of the various slope degradation processes in the Barton Clay Cliffs of Hampshire. *Quarterly Journal of Engineering Geology*, 17, 117–36.

Beardall, C. H., Dryden, R. C. and Holzer, T. J. 1991: *The Suffolk Estuaries*. Suffolk Wildlife Trust.

Bray, M. J. 1996: Beach budget analysis and shingle transport dynamics in West Dorset. Unpublished Ph.D. thesis, London School of Economics, University of London.

Bray, M. J. and Hooke, J. M. 1995: Strategies for conserving dynamic coastal landforms. In M. G. Healy and J. P. Doody (eds), *Directions in European Coastal Management*. Cardigan: Samara Publishing, 275–90.

Bromhead, E. N., Hopper, A. C. and Ibsen, M.-L. 1998: Landslides in the Lower Greensand escarpment in south Kent. *Bulletin of Engineering Geology and Environment*, 57, 131–44.

Brunsden, D. 1991: Coastal and landslide problems in West Dorset. In *Coastal Instability and Development Planning*. SCOPAC, 29–44.

Brunsden, D. and Chandler, J. H. 1996: Development of an episodic landform change model based upon the Black Ven mudslide, 1946–1995. In M. G. Anderson and S. M. Brooks (eds), *Advances in Hillslope Processes*, Vol. 2. Chichester: Wiley, 869–96.

Brunsden, D. and Jones, D. K. C. 1980: Relative time scales and formative events in coastal landslide systems. *Zeitschrift für Geomorphologie Supplementband*, 34, 1–19.

Brunsden, D. and Lee, E. M. in press: Understanding the behaviour of coastal landslide systems: an inter-disciplinary view. In E. N. Bromhead, N. Dixon and M.-L. Ibsen (eds), *Landslides: In Research, Theory and Practice*. London: Thomas Telford.

Bruun, P. 1962: Sea level rise as a cause of shore erosion. *Journal of Waterway, Port, Coastal and Ocean Engineering Division, ASCE*, 88, 117–30.

Burd, F. 1989: *The Saltmarsh Survey of Great Britain: an Inventory of British Saltmarshes*. Research and Survey in Nature Conservation No. 17. Peterborough: Nature Conservancy Council.

Burd, F. 1992: *Erosion and Vegetation Change on the Saltmarshes of Essex and North Kent between 1973 and 1988*. Research and Survey in Nature Conservation No. 42. Peterborough: Nature Conservancy Council.

Cambers, G. 1976: Temporal scales in coastal erosion systems. *Transactions of the Institute of British Geographers*, 1, 246–56.

Camden, W. 1590: *Britannia*. Translated by Gibson.

Carr, A. P. 1969: The growth of Orford Spit: cartographic and historical evidence from the sixteenth century. *Geographical Journal*, 135, 28–39.

Carr, A. P. 1980: *Chesil Beach and Adjacent Area: Outline of Existing Data and Suggestions for Future Research*. Report by the Institute of Oceanographic Sciences, Internal Document 94, to Dorset County Council and Wessex Water Authority.

Carr, A. P. and Blackley, M. W. L. 1974: Ideas on the origin and development of Chesil Beach, Dorset. *Proceedings, Dorset Natural History and Archaeological Society*, 95, 9–17.

Carr, A. P. and Gleason, R. 1972: Chesil Beach, Dorset and the cartographic evidence of Sir John Coode. *Proceedings, Dorset Natural History and Archaeological Society*, 94, 125–31.

Carr, A. P. and Seaward, D. R. 1990: Chesil Beach: changes in crest height, 1965–1990. *Proceedings, Dorset Natural History and Archaeological Society*, 112, 109–12.

Carr, A. P. and Seaward, D. R. 1991: Chesil Beach: landward recession, 1965–1991. *Proceedings, Dorset Natural History and Archaeological Society*, 113, 157–60.

Carter, R. W. G. 1988: *Coastal Environments: an Introduction to the Physical, Ecological and Cultural Systems of Coastlines*. London: Academic Press.

Carter, R. W. G. 1992. Coastal conservation. In M. G. Barrett (ed), *Coastal Zone Planning and Management*. London: Thomas Telford, 21–48.

Carter, R. W. G. and Orford, J. D. 1993: The morphodynamics of coarse clastic beaches and barriers: a short and long term perspective. *Journal of Coastal Research*, 15, 158–79.

Chandler, J. H. 1989: The acquisition of spatial data from archival photographs and their application to geomorphology. Unpublished Ph.D. thesis, City University, London.

Chandler, J. H. and Brunsden, D. 1995: Steady state behaviour of the Black Ven mudslides: the application of archival analytical photogrammetry to studies of landform change. *Earth Surface Processes and Landforms*, 20, 255–75.

Chant, K. 1986: *The History of Dunwich*. Dunwich Museum.

Clark, A. R. and Guest, S. 1991: The Whitby cliff stabilisation and coast protection scheme. In R. J. Chandler (ed.), *Slope Stability Engineering*. London: Thomas Telford, 283–90.

Clark, A. R. and Guest, S. 1994: The design and construction of the Holbeck Hall landslide coast protection and cliff stabilisation emergency works. *Proceedings of the 29th MAFF Conference of River and Coastal Engineers*, 3.3.1–13.

Cobb, R. T. 1957: Shingle Street, Suffolk. *Report of the Field Study Council for 1956–1957*, 31–42.

Codrington, T. 1870: Some remarks on the formation of Chesil Beach. *Geological Magazine*, 7, 23–25.

Collins, K. 1993: Alnmouth: unravelling of an estuary's history. *Geographical Magazine*, December, 53–6.

Conybeare, W. D., Buckland, W. and Dawson, W. 1840: *Ten Plates Comprising a Plan, Sections and Views Representing the Changes Produced on the Coast of East Devon Between Axmouth and Lyme Regis by the Subsidence of the Land . . .* London: John Murray.

Coode, J. 1853: Description of the Chesil Bank, with remarks upon its origin, the causes which have contributed to its formation, and upon the movement of shingle generally. *Minutes of the Proceedings of the Institution of Civil Engineers*, 12, 520–57.

Davidson, N. C., d'A. Laffoley, D., Doody, J. P., Way, L. S., Gordon, J., Key, R., Drake, C. M., Pienkowski, M. W., Mitchell, R. and Duff, K. L. 1991: *Nature Conservation and Estuaries in Great Britain.* Peterborough: Nature Conservancy Council.

Davies, J. 1972: *Geographical Variation in Coastline Development.* London: Oliver and Boyd.

de Boer, G. 1963: Spurn Point and its predecessors. *The Naturalist*, 887, 113–20.

de Boer, G. 1964. Spurn Head: its history and evolution. *Transactions of the Institute of British Geographers*, 34, 71–89.

Derbyshire, E., Page, L. W. F. and Burton, R. 1975: Integrated field mapping of a dynamic land surface: St Mary's Bay, Brixham. In A. D. M. Phillips and B. J. Turton (eds), *Environment, Man and Economic Change.* London: Longman, 48–77.

Doody, J. P. (ed.) 1984: *Spartina anglica in Great Britain.* Focus on Nature Conservation No. 5. Peterborough: Nature Conservancy Council.

Dronkers, J. 1986. Tidal asymmetry and estuarine morphology. *Netherlands Journal of Sea Research*, 20, 117–31.

English Nature 1992: *Campaign for a Living Coast.* Peterborough: English Nature.

Escoffier, E. F. 1940: The stability of tidal inlets. *Shore and Beach*, 1, 114–15.

Fahy, F. M., Hansom, J. D. and Comber, D. P. N. 1993: *Estuaries Management Plans. Coastal Processes and Conservation: Dee Estuary.* Report to English Nature.

Forbes, D. L., Orford, J. D., Carter, R. W. G., Shaw, J. and Jennings, S. C. 1995: Morphodynamic evolution, self-organisation and instability of coarse-clastic barriers on paraglacial coasts. *Marine Geology*, 126, 63–85.

Frew, P. and Guest, S. 1997: Overstrand coast protection scheme. *Proceedings of the MAFF Conference of River and Coastal Engineers*, A.1.1–15.

Gardner, T. 1754: *Historical Notes on Dunwich, Blythburgh and Southwold.* London.

Goa, S. 1993: Sediment dynamics and stability of tidal inlets. Unpublished Ph.D. thesis, University of Southampton.

Goudie, A. S. and Brunsden, D. 1997: *Classic Landform Guide: East Dorset.* Sheffield: The Geographical Association.

Groves, T. B. 1875: The Chesil Bank. *Nature*, 11, 506–7.

Halcrow Group Ltd 2000: *Cowes to Gurnard Coastal Slope Stability Study.* Report to the Isle of Wight Council.

Hansom, J. 1998: The coastal geomorphology of Scotland: understanding sediment budgets for effective coastal management. In *Scotland's Living Coastline.* Edinburgh: Stationery Office, 34–44.

High Point Rendel 2000: *Lyme Regis Environmental Improvements.* Report to West Dorset District Council.

High Point Rendel, in press: *Investigation and Management of Soft Rock Cliffs.* London: Thomas Telford.

Hooke, J. M. and Riley, R. C. 1991: Historical changes on the Hampshire coast 1870–1965. *Proceedings of the Hampshire Field Club and Archaeological Society,* 47, 203–24.

Hutchins, J. 1803: *The History and Antiquities of the County of Dorset.* London: J. Nichols and Son.

Hutchinson, J. N. 1970: A coastal mudflow on the London Clay cliffs at Beltinge, North Kent. *Geotechnique,* 20, 412–38.

Hutchinson, J. N. 1973: The response of London Clay cliffs to differing rates of toe erosion. *Geologia Applicata e Idrogeologia,* 8, 221–39.

Hutchinson, J. N. 1976: Coastal landslides in cliffs of Pleistocene deposits between Cromer and Overstrand, Norfolk, England. In N. Janbu, F. Jorstad and B. Kjaernsli (eds), *Contributions to Soil Mechanics.* Oslo: Norwegian Geotechnical Institute, 155–82.

Hutchinson, J. N. 1991: The landslides forming the South Wight Undercliff. In R. J. Chandler (ed.), *Slope Stability Engineering.* London: Thomas Telford, 157–68.

Hutchinson, J. N. and Gostelow, T. P. 1976: The development of an abandoned cliff in London Clay at Hadleigh, Essex. *Philosophical Transactions of the Royal Society,* A283, 557–604.

Hutchinson, J. N., Poole, C. and Bromhead, E. N. 1985: Combined archaeological and geotechnical investigation of the Roman fort at Lympne, Kent. *Britannia,* XVI, 209–36.

Institute of Estuarine and Coastal Studies (IECS) 1993: *Medway Estuary: Coastal Processes and Conservation.* Report to English Nature, University of Hull, Hull.

Institute of Estuarine and Coastal Studies (IECS) 1994a: *Humber Estuary and Coast.* Humberside County Council, Hull.

Institute of Estuarine and Coastal Studies (IECS) 1994b: *Holderness Coast Defence:* Institute of Estuarine and Coastal Studies, Hull.

Jessop, W. 1805: Report on The Cobb. Lyme Borough Archive, Dorset, RO DC/LR 57.

Jennings, S. C. and Smyth, C. 1990: Holocene evolution of the gravel coastline of East Sussex. *Proceedings of the Geologists Association,* 101, 213–24.

Jennings, S. C. and Orford, J. D. 1999: The Holocene inheritance embedded within contemporary coastal management problems. *Proceedings of the 34th MAFF Conference of River and Coastal Engineers.* London: MAFF, 9.2.1–15.

Jennings, S., Orford, J. D., Canti, M., Devoy, R. J. N. and Straker, V. 1998:

The role of relative sea-level rise and changing sediment supply on Holocene gravel barrier development: the example of Porlock, Somerset, UK. *The Holocene*, 8, 165–81.

Jolliffe, I. P. 1979: *West Bay and the Chesil Bank, Dorset. Coastal Impact of Mining Activities on Coastal Erosion.* Report to West Dorset DC & Dorset CC.

Jones, D. G. and Williams, A. T. 1991: Statistical analysis of factors influencing cliff erosion along a section of the west Wales coast. *Earth Surface Processes and Landforms*, 16, 95–111.

Jones, D. K. C. and Lee, E. M. 1994: *Landsliding in Great Britain.* London: HMSO.

Kidson, C. 1963: The growth of sand and shingle spits across estuaries. *Zeitschrift für Geomorphologie*, 7, 1–21.

Kirby, R. 1990: The sediment budget of the erosional intertidal zone of the Medway estuary. *Proceedings of the Geologists Association*, 101, 63–77.

Lee, E. M. 1998: *The Implications of Future Shoreline Management on Protected Habitats in England and Wales.* Environment Agency R&D Technical Report W150. Bristol: Environment Agency.

Lee, E. M. 1999: Coastal planning and management: the impact of the 1993 Holbeck Hall landslide, Scarborough. *East Midlands Geographer*, 21, 78–91.

Lee, E. M. and Brunsden, D. 1997: *West Bay Geomorphological Study.* Report to West Dorset District Council. Rendel Geotechnics.

Lee, E. M. and Brunsden, D. 1999: *Coastal Processes and Geomorphology: Monmouth Beach.* Report to West Dorset District Council. High Point Rendel.

Lee, E. M. and Moore, R. 1991: *Coastal Landslip Potential Assessment: Isle of Wight Undercliff, Ventnor.* Geomorphological Services Ltd.

Lee, E. M., Moore, R. and McInnes, R. G. 1998: Assessment of the probability of landslide reactivation: Isle of Wight Undercliff, UK. In D. Moore and O. Hungr (eds), *Engineering Geology: the View from the Pacific Rim.* Rotterdam: Balkema, 1315–21.

Leland, J. 1546: The Itinerary of John Leland. T. Hearne (1710–1712).

Lilly, C. 1715: A report on the fortifications and artillery at Portland. *British Museum Kings Geo. III Mss 45, 57.*

Marker, M. E. 1967: The Dee estuary: its progressive silting and saltmarsh development. *Transactions of the Institute of British Geographers*, 241, 65–71.

Matthews, E. R. 1934: *Coast Erosion and Protection.* London: Charles Griffin.

May, V. J. 1971: The retreat of chalk cliffs. *Geographical Journal*, 137, 203–6.

May, V. J. and Heaps, C. 1985: The nature and rates of change on chalk coastlines. *Zeitschrift fur Geomorphologie Supplementband*, 57, 81–94.

Ministry of Agriculture, Fisheries and Food 1994: *Coast Protection Survey of England.* Survey Report, Vol. 1, Summary Report. London: MAFF.

Moore, R., Lee, E. M. and Clark, A. R. 1995: *The Undercliff of the Isle of Wight: a Review of Ground Behaviour.* Ventnor: South Wight Borough Council.

Mottershead, D. 1997: *Classic Landform Guide: South Devon Coast.* Sheffield: The Geographical Association.

O'Brien, M. P. 1931: Estuarine tidal prism related to entrance areas. *Civil Engineering*, 1, 738–9.

Orford, J. D., Carter, R. W. G. and Jennings, S. C. 1996: Control domains and morphological phases in gravel-dominated coastal barriers of Nova Scotia. *Journal of Coastal Research*, 12, 589–604.

Pethick, J. 1992a: Natural change. In M. G. Barrett (ed.), *Coastal Zone Planning and Management*. London: Thomas Telford, 49–63.

Pethick, J. 1992b: Salt marsh geomorphology. In J. R. L. Allen and K. Pye (eds), *Saltmarshes: Morphodynamics, Conservation and Engineering Significance*. Cambridge: Cambridge University Press, 41–62.

Pethick, J. 1994: Estuaries and wetlands: function and form. In *Wetland Management*. London: Thomas Telford, 75–87.

Pethick, J. 1995: Estuarine processes. In National Rivers Authority, *A Guide to the Understanding and Management of Saltmarshes*. Bristol: NRA, 33–72.

Pethick, J. 1996a: Coastal slope development: temporal and spatial periodicity in the Holderness Cliff Recession. In M. G. Anderson and S. M. Brooks (eds), *Advances in Hillslope Processes*, Vol. 2. Chichester: Wiley, 897–917.

Pethick, J. 1996b: The geomorphology of mudflats. In: K. F. Nordstrom and C. T. Roman (eds), *Estuarine Shores: Evolution, Environments and Human Alterations*. Chichester: Wiley, 185–211.

Pethick, J. 2000: Choosing the future: managing coastal systems. Workshop paper, University of Newcastle.

Pitts, J. 1983: The temporal and spatial development of landslides in the Axmouth–Lyme Regis Undercliff, National Nature Reserve, Devon. *Earth Surface Processes and Landforms*, 8, 589–603.

Posford Duvivier, 1990: *Lyme Regis Environmental Improvements: a Coast Protection and Harbour Management Scheme – Consulting Engineer's Appraisal*.

Posford Duvivier 1993: *Report on Proposals for Coast Protection at Easington, Holderness Borough Council*.

Pye, K. 1996: Evolution of the shoreline of the Dee estuary, United Kingdom. In K. F. Nordstrom and C. T. Roman (eds), *Estuarine Shores: Evolution, Environments and Human Alterations*. Chichester: Wiley, 15–37.

Pye, K. and French, P. W. 1992. *Targets for Coastal Habitat Recreation*. Science Report 35. Peterborough: English Nature.

Pye, K. and French, P. W. 1993: *Erosion and Accretion Processes on British Saltmarshes*. Report to Minstry of Agriculture, Fisheries and Food.

Rampling, P. A., Jago, C. F. and Scourse, J. D. 1996: Contemporary and Holocene sediment dynamics of the Walton Backwaters, North-east Essex. In NRA, National Rivers Authority, *Saltmarsh Management for Flood Defence*. Bristol: 34–45.

Redman, J. B. 1852: On the alluvial formations, and the local changes, of the south coast of England. *Minutes of the Proceedings of the Institution of Civil Engineers*, 11, 162–223.

Rendel Geotechnics 1995: *Applied Earth Science Mapping: Seaham to Teesmouth*. Department of the Environment.

Robinson, A. H. W. 1955: The harbour entrances of Poole, Christchurch and Pagham. *Geographical Journal*, 121, 33–50.

Robinson, D. A. and Williams, R. B. G. 1984: *Classic Landforms of the Weald.* Sheffield: The Geographical Association.

Rozier, I. T. and Reeves, M. J. 1979: Ground movements at Runswick Bay, North Yorkshire. *Earth Surface Processes and Landforms*, 4, 275–80.

Saiu, E. M. and McManus, J. 1998: Impacts of coal mining on coastal stability in Fife. In J. M. Hooke (ed.), *Coastal Defence and Earth Science Conservation.* London: Geological Society, 58–66.

Schofield, J. 1787: *An Historical and Descriptive Guide to Scarborough and its Environs.* York: W. Blanchard.

Searle, S. A. 1975: *The Tidal Threat: East Head Spit, Chichester Harbour.* Chichester Harbour Conservancy, Chichester.

Simm, J. D., Brampton, A. H., Beech, N. W. and Brooke, J. S. 1996: *Beach Management Manual.* Report 153. London: CIRIA.

Sims, P. and Ternan, L. 1988: Coastal erosion: protection and planning in relation to public policies – a case study from Downderry, south-east Cornwall. In J. M. Hooke (ed.), *Geomorphology in Environmental Planning.* Chichester: Wiley, 231–44.

So, C. L. 1967: Some coastal changes between Whitstable and Reculver, Kent. *Proceedings of the Geologists Association*, 77, 475–90.

Steers, J. A. 1946: *The Coastline of England and Wales.* Cambridge: Cambridge University Press.

Steers, J. A. 1951: Notes on erosion along the coast of Suffolk. *Geological Magazine*, 88, 435–9.

University of Strathclyde 1991: The assessment and integrated management of coastal cliff systems. Report to Ministry of Agriculture, Fisheries and Food.

Valentin, H. 1954: Der landverlust in Holderness, Ostengland von 1852–1952. *Erde*, 6, 296–315.

Webber, N. B. 1979: *An Investigation of the Dredging in Chichester Harbour Approach Channel and the Possible Effects on the Hayling Island Coastline.* University of Southampton.

Williams, A. T. and Davies, P. 1987: Rates and mechanics of coastal cliff erosion in Lower Lias rocks. *Proceedings of Coastal Sediments '87*, 1855–70.

Williams, A. T., Morgan, N. R. and Davies, P. 1991: Recession of the littoral zone cliffs of the Bristol Channel, UK. In O. T. Morgan (ed.), *Coastal Zone '91*, 2394–408.

Wright, L. D., Coleman, J. M. and Thom, B. G. 1973: Processes of channel development in a high-tide range environment: Cambridge Gulf–Ord River Delta, W. Australia. *Journal of Geology*, 81, 15–41.

Young, Rev. G. and Bird, J. 1822: *A Geological Survey of the Yorkshire Coast.* Whitby.

Chapter 7

Sediment Transfer in Upland Environments

*David L. Higgitt, Jeff Warburton
and Martin G. Evans*

While the outlines of the lowlands are touched with the instability that
marks everything human, these far heights seem to remain impassive and
unaffected, as if the hand of time had passed them by. Hence the
everlasting hills have ever been favourite emblems, not only of grandeur
but of immutable permanence. And yet the mountains bear on their
fronts the memorials of change which have not altogether failed to catch
the eye ... These grim, naked cliffs and splintered precipices, their
yawning defiles and heaps of ruins, have always appealed to the fancy
and fear of men.

Archibald Geikie, *The Scenery of Scotland* (2nd edn), 1887, p. 1.

7.1 Introduction

The impact of glacial activity dominates the geomorphology of upland
Britain. In addition to the sculpting of classic glacial scenery, the
transport and deposition of the products of glacial erosion provide a
substantial legacy that has conditioned-postglacial geomorphological
processes. And yet across many British mountains, glacial erosion has
only partly modified topographic surfaces and drainage networks that
were created in the Tertiary. In consequence, much of the upland
landscape is inherited from past conditions that provoke Geikie's
contemplation of immutable permanence. A useful analogy is landscape
as palimpsest – a term for a medieval parchment on which old text has
been partly removed to make way for new. The purpose of this chapter
is to consider the most recent additions to a landscape text dominantly
drafted by glacial activity. How much of the present upland landscape
can be attributed to changes during the last millennium, and what

evidence is there for enhanced processes of sediment transfer over this timescale? There is a growing interest in late Holocene landscape changes in upland Britain, much of which has been propelled by investigations of valley-floor alluviation and recognition of their ecological significance (Ratcliffe and Thompson, 1988). Evidence for enhanced erosion on upland slopes during the late Holocene has been little documented and is often hampered by a lack of precise dating (Gerrard, 1991). An assessment of upland sediment transfer should therefore begin with consideration of the reasons for the limited extent of documented evidence about changing process regimes in the last millennium.

Geomorphological change during the last 1000 years may not be known because of the limited detailed study undertaken thus far. Much progress has been made in the last 20–30 years in recognizing late Holocene slope and fluvial processes (Macklin et al., 1992), but many parts of upland Britain have not been investigated in detail. Improved coverage and the development of techniques may supplement current knowledge. Second, evidence of the past extent of process activity may not be preserved in the landscape. This is illustrated by the case of peat slides (discussed later in this chapter, in section 7.2.3). A number of failures have been described in the literature or in newspaper reports, but the recovery of vegetation makes recognition of these features difficult within 50–100 years. Third, the extent of landscape change in the last 1000 years is compounded by the difficulties of dating change precisely. Historical documents have proved useful for reconstructing climate and noting the occurrence of particular storms, but rarely describe geomorphological change, especially in the uplands. Radiometric techniques measuring caesium-137 and lead-210 content in depositional sequences permit dating back to about 50 and 150 years respectively. Carbon-14 dates are generally applicable to sediments older than about 300 years, but are complicated where the upstream sediment supply includes peat. Interpretation is therefore constrained by a dating gap at a time when both climate change and land use change was likely to have been significant. Dating is also compromised by a lack of suitable material. Depositional sequences may lack the organic matter required for ^{14}C dating, while dating of erosional phases necessitates interpretation of proxy data. A consequence of the lack of precision in dating means that linking erosion to either climatic deterioration or vegetation degradation is largely circumstantial (Ballantyne, 1991). Untangling the relative importance of human-induced, climate-driven or extreme events remains elusive and largely open to spurious correlation. The final reason for limited published research might reflect the lack of significant change and relative unimportance of the last millennium in the uplands. Although land use change has been significant during the last 1000 years, palynological reconstructions indicate that regional forest clearance was well established before the onset of the last millennium, as noted previously in chapter 3. A framework for considering climate and land use change in the uplands over the last 1000 years is provided in figure 7.1, which is based on the experience of northern England (North Pennines and Lake District). Circumstances for other upland regions vary in detail. Working at this timescale presents a challenge to geomorphology, which has become accustomed to considering the contemporary or the longer Holocene timescale (Lewin, 1980; Roberts, 1998).

Bearing in mind the limitations of the recognition, preservation and dating of upland slope processes during the last millennium, the main objective of the chapter is to review and evaluate the evidence for process activity in the British uplands. The sensitivity of peat and soils to erosion is examined before considering the role of mass movement and periglacial activity during the last 1000 years. Most studies of upland erosion have concentrated on water-related processes and the potential of wind in sediment transfer may have been overlooked. Many British uplands have been profoundly affected by mining, particularly in the last 200 years, and the geomorphological consequences are summarized. Most recently, recreational pressures have also started to affect the stability of upland systems (Thompson et al., 1987). The wider impact of sediment mobility on the geomorphological system depends on the coupling between slope and channel systems. Millennial-scale change in river channels and floodplains is discussed in chapter 4, but the application of sediment budgets is considered briefly here.

7.2 Upland Slope Processes

7.2.1 Peat erosion

Blanket peat covers 8% of the land surface of Great Britain, representing the single largest upland cover type. In global terms, the blanket mire ecosystems of the British uplands are of international significance. In many areas, the peat is actively eroding or shows signs of significant past erosion (Bower, 1960, 1961; Tomlinson, 1981; Tallis, 1998). Dating the phases of erosion is difficult, not only in attempting to date something that has been removed, but there is the additional problem that fluvial deposits downstream of peatlands include preserved organics such that ^{14}C dating may be unreliable. Attempts to date phases of

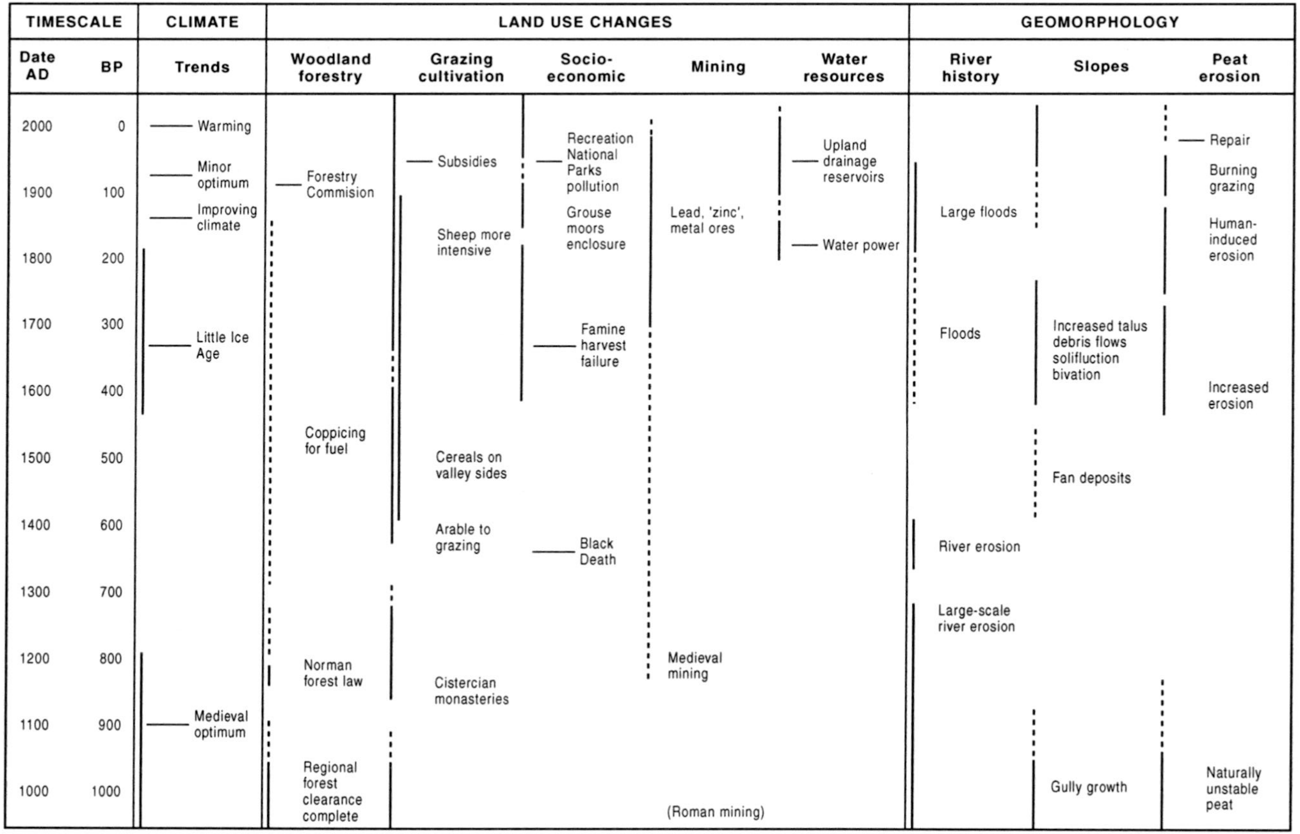

Figure 7.1 A schematic chronology of landscape change in the uplands of northern England in the last 1000 years. The land use changes are based on the experience of the northern Pennines and the Lake District, and can be applied to other upland areas with some modification.

TIMESCALE		CLIMATE	LAND USE CHANGES					GEOMORPHOLOGY		
Date AD	BP	Trends	Woodland forestry	Grazing cultivation	Socio-economic	Mining	Water resources	River history	Slopes	Peat erosion
2000	0	Warming		Subsidies	Recreation National Parks pollution		Upland drainage reservoirs			Repair
1900	100	Minor optimum	Forestry Commission							Burning grazing
1800	200	Improving climate		Sheep more intensive	Grouse moors enclosure	Lead, 'zinc', metal ores	Water power	Large floods		Human-induced erosion
1700	300	Little Ice Age								
1600	400				Famine harvest failure			Floods	Increased talus debris flows solifluction bivation	Increased erosion
1500	500		Coppicing for fuel	Cereals on valley sides					Fan deposits	
1400	600			Arable to grazing	Black Death			River erosion		
1300	700							Large-scale river erosion		
1200	800		Norman forest law	Cistercian monasteries		Medieval mining				
1100	900	Medieval optimum	Regional forest clearance complete							
1000	1000					(Roman mining)			Gully growth	Naturally unstable peat

peat erosion have tended to use evidence from lake cores or to link instability to vegetation change which can be inferred from the pollen record (Conway, 1954; Ashmore et al., 2000). In many ways the sensitivity of blanket peat is analogous to badland erosion (Tallis, 1997a). The highly erodible substrate is protected by the vegetation community that it supports. The vegetation community might be placed under pressure by climatic changes, including increased storminess or desiccation during prolonged dry periods, or by fire, overgrazing and pollution (Birks, 1988; Tallis, 1997a). Degradation of the vegetation community may lead to erosion by crossing of an extrinsic threshold. Peat erosion may also be initiated by the crossing of intrinsic thresholds, such as the expansion of blanket peat on to steeper slopes and consequent mass failure.

Some examples of studies that have attempted to date phases of peat erosion are presented in figure 7.2 where the distribution of blanket peat in the British Isles is illustrated. Step changes in organic content from seven lakes across Britain and Ireland indicate the initiation of peat erosion in the catchment (Rhodes and Stevenson, 1997). Each core has been dated using ^{210}Pb supported by ^{14}C. Initiation dates for different lakes range from AD 900 to 1800, but cluster at 1500–1700, which the authors infer is due to increased storminess of the Little Ice Age (figure 7.2(b)). Bradshaw and McGee (1988) use a similar approach, supported by ^{14}C chronology, for two Irish lakes: Art Lough in the south and Lough Nackbraddy in the north. Peat contribution to lake sediment increases from c. AD 500. In this case, the authors suggest that erosion results from the crossing of intrinsic thresholds, considering the accumulation and erosion of blanket peat to be a cyclical phenomenon. In the southern Pennines, Tallis has worked extensively on dating peat erosion episodes using intact peat stratigraphy. Pollen and macrofossils can be used to establish the relative wetness of the surface at different periods of time. *Racomitrium* moss remains in uneroded profiles signify drier periods but it is continuously present in hummock sites following the initiation of gully erosion because of more effective drainage (Tallis, 1995). The onset of gullying is indicated by a decline in *Sphagnum* and increases in *Racomitrium* and *Empetrum* (figure 7.2(c)). These species indicate drier bog surface conditions due to drainage associated with the gullying (Tallis, 1997b). The modal dates for initiation in the southern Pennines are AD 1250–1450, suggesting that the onset of erosion at this time was linked to desiccation of the peat surface in the Medieval Warm Period.

More recent phases of peat erosion have been linked to human modification of vegetation, in particular overgrazing (Shimwell, 1974), and the impact of air pollution. *Sphagnum* is particularly susceptible to

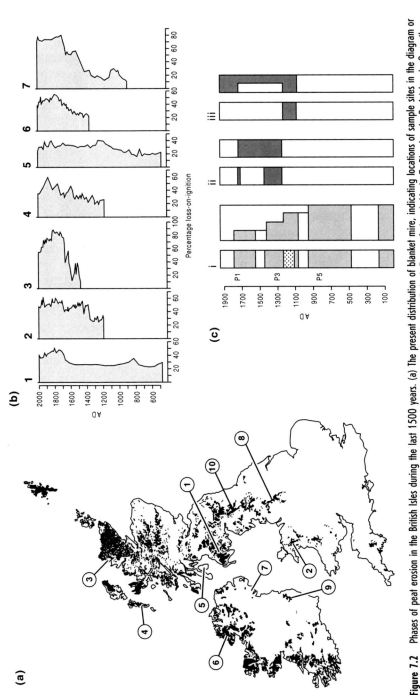

Figure 7.2 Phases of peat erosion in the British Isles during the last 1500 years. (a) The present distribution of blanket mire, indicating locations of sample sites in the diagram or mentioned in the text (1, Round Loch of Glenhead; 2, Llyn Conwy; 3, Loch Na Larch; 4, Loch Teanga; 5, Loch Tanna; 6, Lough Muck; 7, Blue Lough; 8, Art Lough; 9, southern Pennines; 10, Moor House NNR). (b) Profiles of percentage loss-on-ignition of lake sediment cores from sites 1–7 (after Rhodes and Stevenson, 1997). (c) A comparison of southern Pennine macrofossil stratigraphy of (i) *Sphagnum* and (ii) *Racomitrium lanuginosum* and the pollen stratigraphy of *Empetrum* (after Tallis, 1997b). The left-hand column of each pair represents uneroded sites and the right-hand column eroded sites.

SO_2 and there is some indication that recent episodes of destabilization of the blanket peat coincided with the Industrial Revolution, particularly in the southern Pennines, which is adjacent to the conurbations of Manchester and Sheffield. However, the evidence is far from universal. In southern Scotland, incorporation of organics signifying peat erosion predates industrial pollution by at least 100 years and more likely represents attempted land improvement (Stevenson et al., 1990). In some parts of the southern Pennines, *Sphagnum* was lost 400–500 years ago due to reduced surface wetness rather than air pollution (Tallis, 1994), while Mackay and Tallis (1996) interpret the decline in *Sphagnum* and consequent erosion on part of the Forest of Bowland to a combination of drought and a catastrophic burn.

Reliance on temporal correlation to infer causes of peat erosion is clearly problematic because of the variety of factors that could account for its initiation. Locally, it is possible to see evidence for each of these mechanisms in a moorland landscape, and to identify parts of the peat blanket that are actively eroding while other parts are revegetating. Recent revegetation of gully floors is evident in the Moor House National Nature Reserve, northern Pennines (figure 7.3). Some regional variations are also apparent. Whereas the gullied pool and hummock topography of southern Pennine blanket peat is often fringed by small slumps, the northern Pennines have less extensive gully networks but support more dramatic mass failures.

7.2.2   Erosion of mineral soils

Soil erosion has been identified as a serious issue in some upland areas (Royal Commission on Environmental Pollution, 1996). The paucity of data on the extent, frequency and magnitude of erosion makes it difficult to establish trends at the present time, let alone across the last 1000 years. Attempts to develop baseline information include the mapping of erosional features in a sample of aerial photographs of the Scottish Highlands (Grieve et al., 1995). Supported by Scottish Natural Heritage, the study provides quantitative information for the occurrence of different erosional features within physiographic regions or altitude classes. A similar initiative for England and Wales has been supported by the Ministry of Agriculture, Fisheries and Food. Such surveys may prove valuable for providing the benchmark to assess future erosion in the uplands, but the question remains about the significance of present-day rates relative to the past. Areas of bare soil provide few clues of their initiation and growth. An 11 year sequence of photographs of erosion scars on Derwent Edge in the Peak District

Figure 7.3 The revegetation and stabilization of gullied peat over a 40 year period, Burnt Hill, Moor House National Nature Reserve, northern Pennines. (a) 1958; (b) 1998.

(Evans, 1989) demonstrates the expansion of scar area at an average of 4% per year, which is attributed to overgrazing by sheep.

Numbers of sheep have increased significantly in most parts of Britain during the twentieth century, doubling to 44 million between the 1940s and 1993 (Sansom, 1998). This is five times the population compared to the 1860s, although the Highland Clearances in Scotland led to dramatic increases in stocking densities in the nineteenth century. Erosion sensitivity is enhanced through both the damage to the protective vegetation cover and the compaction of soils by trampling (Evans, 1998). Paired run-off studies on improved pasture on the Clwydian Hills of North Wales demonstrate enhanced overland flow on grazed surfaces (James and Alexander, 1998). In the Peak District a decline in heather and bilberry cover and its replacement by grassland throughout the twentieth century is coincident with a decline in management for grouse habitat and with an increase in sheep numbers (Evans, 1989). More specifically, numbers of sheep have responded to the introduction of subsidy legislation, in the case of the Peak District to the 1947 Agriculture Act. Similar observations of a correlation between vegetation degradation and sheep numbers have been reported from many British upland areas (Sansom, 1996) and there is tentative evidence for flashier run-off response. However, it remains difficult clearly to separate the impact of sheep from other factors that might impact on run off and sediment transfer such as moorland drainage, changing seasonal distribution of rainfall or the impacts of other animals, such as fluctuating rabbit populations. Furthermore, it is impossible to establish the distribution of exposed soils and sheep scars beyond the photographic record. Evidence for older phases of soil erosion or slope instability must be inferred from dating restabilized surfaces or sediments that have accumulated downslope.

7.2.3 Slope instability

Lichenometry provides a means of dating the length of time elapsed since a rock- or debris-covered slope stabilized, provided that a reliable lichen growth curve can be derived for the field area. Innes (1983, 1989) applied lichenometry to 780 debris flow deposits in the Scottish Highlands (see also figure 2.10(c)). His interpretation suggests that there were few events prior to 1700 and that the majority of debris flows have occurred in the last 250 years coinciding with land use change in the Highlands, particularly a sharp increase in sheep grazing densities in the nineteenth century. It is notable that the timing of Scottish debris flows postdates the main climatic deterioration of the Little Ice Age compared with the recorded frequency of mass movements in southern Norway (Grove, 1972) which peaked between AD 1650 and 1750. The interpretation rests on the assumption that the burial of pre-existing deposits by each subsequent debris flow was minimal.

The relationship between alluvial fans, debris cones and terrace sequences in upland valleys provides clues to the timing of slope instability. The periglacial environment and limited vegetation cover of the Late Glacial promoted active slope processes and the modification of drift-covered valley-side slopes by solifluction processes. Reduction of sediment supply in the Holocene led to a dominant trend of valley incision that has been episodic and has formed sequences of river terraces (Harvey, 1985; and discussed in chapter 4). Superimposed on this sequence of postglacial valley development are indications of active slope processes during the last millennium. In the Howgill Fells of Cumbria, reactivation of slope processes can be dated from organic layers in terrace deposits buried by colluvial deposits. The deposition of a debris cone over the low terrace of Langdale valley (Harvey et al., 1981) terminates a stable period which persisted for about 1500 years until 940 ± 95 BP, which is associated with Viking settlement in the area. Similar dates for renewed activity have been reported in other parts of northwestern England and southern Scotland. Viking settlement is also inferred from eleventh-century cone development in Langden Valley in the Forest of Bowland (Harvey and Renwick, 1987). In Yarrow Water, Southern Uplands, a palaeosol buried by renewed activity on the debris cone of Dry Cleuch, gives an upper date of 895 ± 45 BP (Dunsford, 1998), while Tipping and Halliday (1994) deduce eleventh- to twelfth-century dates for aggradation of a tributary fan and a low alluvial terrace in the upper Tweed valley. Although the similarity of dates for the fan and terrace are apparent, the stratigraphic relations suggest that whatever destabilized the tributary catchment causing fan aggradation did not simultaneously affect the larger Tweed system.

Further to the north, where agricultural disturbance occurred later, the chronology of fan development is slightly different. In Glen Etive a similar pattern of paraglacial aggradation of debris cones is observed between 10 and 4 k BP, stabilizing as sediment supplies from gullies became exhausted (Brazier et al., 1988). This implies that paraglacial adjustment had been completed before the start of the last millennium and that the system had become weathering-limited. Soil development on the cone surface ended at 550 ± 50 BP with the destruction of the vegetation cover, fluvial incision and the development of an alluvial fan. In this case, agricultural activity and associated burning of the cone surface itself appears to be responsible for erosion, rather than enhanced sediment supply from the gullies above. The most recent phase of slope processes recorded in a debris cone in Glen Docherty is dated to 450 cal. BP and may reflect local woodland clearance or extreme rainfall events during the Little Ice Age (Curry, 2000). However, mid-Holocene phases of activity at this site point to discrete local storms as causal factors rather than long-term climate change. A section of an exposed late Holocene fan sequence in the Grampian Highlands is also interpreted as comprising three hyperconcentrated flows generated by individual storms of short duration, the most recent occurring between 900 and 700 cal. BP (Ballantyne and Whittington, 1999). The implication is that episodes of sediment accumulation triggered by extreme events are independent of either large-scale anthropogenic or climate control and, as noted earlier by Tipping and Halliday (1994), the temptation to infer contemporaneity between different sites may be misleading. Harvey (1986) has evaluated the geomorphological impact of a 100 year event in the Howgills. Peak rainfall intensities were sufficient to trigger shallow turf slides on slopes, to reactivate stabilized gully slopes and to entrench gully beds. The shallow landslides were not coupled with the drainage network, but the gully systems delivered substantial amounts of material to channels that showed little evidence of major sediment input, since the phase of instability associated with Viking settlement. A convectional storm over the Forest of Bowland in 1967 also triggered shallow slope failures (Newson and Bathurst, 1990), many of which remain unvegetated. It is clear that exceptional events can cause substantial geomorphological change in upland environments and their role relative to progressive climate change and land use change should be re-evaluated.

Compared to northern and north-west Britain, no evidence of late Holocene alluvial fan sedimentation has been reported for the Cheviots, Pennines or Welsh uplands. However, the lowest two terraces in Upper Wharfedale in the Yorkshire Dales have recently been attributed to phases of aggradation in the seventh and eleventh centuries AD, respectively (Howard et al., 2000). The lower terraces are composed of markedly finer material than the two higher early Holocene terraces, which is attributed to an input of soil from agricultural expansion. The lowest unit marking the onset of the last millennium is also contaminated with heavy metals, which may reflect early mining activity. In the Cheviots, there is evidence for accelerated fluvial activity that is temporally correlated with the Little Ice Age, followed by marked incision from the eighteenth century (Tipping, 1994).

Mass movements of peat, normally reported as peat slides and bog bursts, have been well documented for over the last 150 years, yet the fundamental controls of this form of shallow instability are still poorly understood. Reports of peat slides, bog flows and bursts are widespread in the uplands of the British Isles (Crisp et al., 1964; Tomlinson and Gardiner, 1982; Alexander et al., 1986). Most instabilities are associated with convective summer thunderstorms, although instances of rapid snowmelt and intense winter rainfall have also triggered these events. Peat slides are important because they mobilize considerable quantities of surficial peat; they have major impacts on stream ecosystems; and present a local hazard to humans. For example, a slide in

Baldersdale in Northern England in 1689 caused considerable flood damage and poisoned fish for several kilometres downstream (Archer, 1992). The clustering of peat slides in recent centuries may reflect human interference in peatlands coupled with periods of increased storminess. Earlier peat slides certainly exist, but the evidence for these is either not recorded or masked by the regrowth of vegetation and degradation of failure scars in the field. Discontinuities in peat cores are now being recognized as indicators of instability on adjacent hillslopes, which offers opportunities for dating much older peat slides (Ashmore et al., 2000).

7.3   Periglacial Activity

Following deglaciation, periglacial processes substantially modified and redistributed material but diminished in effectiveness as vegetation cover established, sediment supplies exhausted and climate warmed. Although modest in comparison because of limited frost penetration, there is some evidence for some rejuvenation of periglacial processes in the last millennium (Sugden, 1971). The extreme wetness and strong winds that characterize contemporary conditions on Scottish mountains have been termed a 'maritime periglacial regime' (Ballantyne and Harris, 1994). Within the last millennium, enhanced frost activity would have occurred when climatic conditions increased the number of freeze–thaw cycles. Reconstructed winter temperature series do not provide the necessary detail to ascertain precisely how frost heave and needle ice formation would have accelerated geomorphological activity through the Little Ice Age. A number of field and experimental studies have demonstrated the effectiveness of needle ice in weakening exposed sediment, such as exposed peat or river banks (Lawler, 1993). On upland hillslopes needle ice will extend scars initiated by sheep and retard seedling establishment on exposed soils (Legg et al., 1992) and it seems reasonable to assume that during colder phases of the last millennium many drift-covered slopes would have been potentially unstable. Solifluction involves a combination of frost heave and local liquefaction that produces a slow downslope transfer of surface material as sheets or lobes and as such can be accelerated through changes in near-surface hydrological conditions as well as by frost activity. Radio-carbon dates on a soil horizon buried by the downslope advance of a solifluction lobe near the summit of the Fannich Mountains in north-west Scotland indicate an earliest date for advance between 890 ± 120 to 590 ± 90 BP (Ballantyne, 1986). Whether this recent activity has been triggered by climate deterioration of the Little Ice Age or through degradation of the vegetation cover by overgrazing is unclear. Late Holocene solifluction features can be distinguished from older counterparts by their smaller size, lack of sorting and steeper risers (Ballantyne, 1987). Two populations of solifluction terraces can be observed on Lake District mountains, such as Skiddaw, and though the smaller subset has not been dated.

Patterned ground – the formation of stone polygons and stripes by frost heave – is a characteristic of periglacial activity in many upland areas. Some studies report active contemporary development of stone sorting (Caine, 1963; Warburton, 1987; Ballantyne, 1996; Warburton and Caine, 1999), but a more general trend of the degradation and disappearance of relict features. Although a geomorphological curiosity, patterned ground has limited relevance to sediment transfer except to note that frost heave on exposed surfaces can liberate substantial quantities of material. This can be a particular problem on bare, exposed footpaths (Bayfield et al., 1988). The role of wind in redistributing material on exposed upland surfaces, frequently associated with present-day cold environments, is of greater significance but has received relatively limited attention (King, 1971; Ball and Goodier, 1974; Goodier and Ball, 1975).

7.4 Aeolian Processes

Mantles of aeolian sediment have been noted on mountain summits in Scotland (Pye and Paine, 1983; Ballantyne, 1998) and Ireland (Wilson, 1989). Deposits reaching almost 3 m in depth on The Storr, Isle of Skye, accumulated following exposure of the present rockwall by a landslide around 6500 ± 500 cal. BP. Dated palaeosols within the sequence indicate considerable accumulation in the last 3000 years, although it is difficult to infer rates specifically for the last millennium. Nearby on the Scottish mainland, prolonged accumulation of aeolian sands on An Teallach was terminated by disruption of the vegetation cover followed by deflation and redistribution to lee slopes during the seventeenth and eighteenth centuries (Ballantyne and Whittington, 1987). The destruction of the vegetation cover might reflect increase storminess during the Little Ice Age or the introduction of sheep to high ground. Active wind-blown redistribution from exposed windward to lee slopes can be observed on the ultrabasic plateau of the Isle of Rum.

Aeolian sediment transport is also evident on heavily eroded blanket peat topography. In extreme cases, such as the Forest of Bowland, the remnant peat islands are surrounded by a lag surface from which the majority of peat has been deflated. There appear to have been no attempts to quantify wind-blown sediment flux on British blanket peat, although the role of deflation on remnant peat has been noted (Ball and Goodier, 1974) and, more recently, reported in Finland (Luoto and Seppala, 2000).

7.5 Mining Impacts

Many British upland areas have been impacted by mining activity (Macklin and Rose, 1986). In some localities production from Roman times can be detected in the landscape, but it was expansion during medieval times and in the eighteenth and nineteenth centuries that had the most significant geomorphological impact (Macklin, 1999). Sediments released by mining continue to have a major influence on valley-floor alluviation downstream, and are considered in more detail in chapter 4. Heavy metal concentrations that can be related to records of ore production can be used as a proxy to date terrace and development (Hudson-Edwards et al., 1999; Macklin et al., 2000) and provide a source of contaminated material for remobilization (Rowan et al., 1995b). On upland slopes sediment transfer is high, albeit locally concentrated. The exploration technique known as 'hushing' involved damming and releasing small streams to strip away the drift cover and expose mineral veins. Based on estimates of rock removal from the North Pennines ore field, average sediment yields during the main mining phases in the nineteenth century would have exceeded 200 t km^{-2} yr^{-1}, about ten times the present stream loads. Much of this may have been deposited in the vicinity of the mines and was not necessarily coupled to the stream system. On the other hand, the estimate of rock removal excludes sediment mobilized by hushing.

Periods of increased storminess associated with the Little Ice Age may be responsible for increasing coupling between hillslope sediment stores and stream channels (Hudson-Edwards et al., 1999). Such periods of instability increase sediment delivery to stream channels which respond to adjustments in channel form (Macklin et al., 1998). The linkage between hillslope instability and channel storage remains unclear partly due to the lack of knowledge of the chronology of hillslope sediment delivery and partly due to the complex response of individual river systems to the imposed sediment load. Merrett and Macklin (1999) present historical evidence from the Yorkshire Dales which clearly demonstrates that in catchments where sediment supply was augmented by metal mining activity, channel incision in the late nineteenth century was delayed.

7.6   Upland Sediment Budgets

A sediment budget is a procedure to account for the production, transfer and export of sediment from a given catchment area. Conceptually, it is valuable as a means of identifying the relative importance of particular processes at different hydrological conditions and in articulating the linkages and pathways of transfer. It is well known that the amount of sediment exported from a catchment is usually only a small fraction of the gross erosion (Walling, 1983). As catchment area increases, the ratio of yield to gross erosion (the sediment delivery ratio) tends to decrease, although this relationship has been observed mainly in agricultural systems that are transport-limited. Given that sediment transport in some headwater catchments may have become weathering-limited following the exhaustion of paraglacial sediment supply, it is possible that the sediment delivery may increase downstream, as has been observed in the glacially scoured river basins of western Canada (Church et al., 1989). Either way, catchment scale and sediment delivery has implications for how observed or reconstructed sediment yield at a point can be related to the geomorphological activity in the catchment, such that a full sediment budget would account for rates and processes of erosion on hillslopes, sediment transport in the fluvial network, temporary storage of material in alluvial fans, channel bars and floodplains and for the breakdown of sediments in transport or storage (Reid and Dunne, 1996). While most attempts to construct sediment budgets have focused on describing contemporary patterns of sediment transfer, the approach is suitable reconstructing the sensitivity of systems to past conditions, in particular the consequences of climate and land use change.

Lakes and reservoirs provide an opportunity to estimate historical sediment yields. Radiometric techniques supported by proxy indicators such as mineral magnetics, heavy metals and geochemistry allow a chronology of sedimentation to be established. The abundance of reservoirs in many parts of upland Britain, from the mid-nineteenth century onwards, affords information on local and regional variation of sediment yields for the last 100–150 years. The geographical distribution of lakes is more limited and the resolution of the bottom sediments is poorer, but a sediment budget approach can be extended to the millennial scale. Reviewing the value of lakes and reservoirs for geomorphological studies, Dearing and Foster (1993) emphasize how the complexity of lake sedimentation may impinge on sediment budget interpretation. Estimation of catchment sediment yield is constrained by sediment delivery processes, within-lake sedimentation patterns –

remobilization, trap efficiency, and by operational and methodological procedures. The scale dependence of the specific sediment yield is augmented by an observed inverse relationship with the catchment to lake ratio (Foster et al., 1990). Higher sediment yields are found where catchment areas are small relative to lake area, explained by the connectivity between slopes, channels and the lake basin and by limited storage area. As catchment to lake ratios increase, fluvial transport becomes more significant in sediment transfer mechanisms, from which it follows that, other things being equal, small lake basins are more likely to reflect slope processes directly. Sedimentation patterns within lakes vary over time, requiring multiple cores to obtain reliable data on sediment accumulation over time. Similarly, substantial areas of lake and reservoir beds in upland regions slope steeply and are not active accumulation zones. In reservoirs, as the water level is drawn down, sediment sequences can be incised and reworked down basin. Estimation of catchment erosion rates is also constrained by the trap efficiency of the lake, itself a function of capacity relative to water inflow (Brune, 1953). In rapidly accumulating reservoirs, trap efficiency declines with time and must be accounted for in the model of sediment yield variability. Reservoir operating procedures, including water transfers, dredging and changing outflow level, also impact on the sediments preserved. Finally, information obtained from upland lakes and reservoirs is constrained by the methods employed. Detailed investigations based upon the correlation of multiple cores are costly to undertake and have been limited to a few sites. Interpretation of geomorphological activity over the last 1000 years using lake sediments is clearly not as straightforward as it might first appear.

There have been a number of attempts to summarize regional or national trends in sediment yield from existing reservoir surveys (Walling and Webb, 1981; Butcher et al., 1993; Duck and McManus, 1994). Walling and Webb (1981) suggest that a sediment yield of 100 t km^{-2} yr^{-1} is typical of upland environments, but even within local areas large differences are apparent. Estimated sediment yields for 28 southern Pennine reservoirs (Butcher et al., 1993) range over two orders of magnitude from 2.9 to 289 t km^{-2} yr^{-1}. Employing seven bathymetric surveys, Rowan et al. (1995a) have been able to calculate variations in sediment yield to Abbeystead Reservoir, Forest of Bowland. The 140 year average sediment yield is 192 t km^{-2} yr^{-1}, but decadal-scale variations range from 78 to 390 t km^{-2} yr^{-1}. The highest yields occurred between 1930 and 1948, which coincides with drainage and improvement of upland grazing. Foster and Lees (1999) report century average sediment yields between 9 and 35 t km^{-2} yr^{-1} for four upland reservoirs in North-East England. To underline the problems of estimating sediment yields, their estimate for Silsden Reservoir, West Yorkshire, based on multiple cores is an order of magnitude lower than one based on bathymetric survey by Butcher et al. (1993).

Reconstruction of past sediment yields can be extended further back using natural lake basins rather than reservoirs, subject to their more limited geographical distribution. In North Wales, Dearing (1992) reports a mean suspended sediment yield between 6 and 18 t km^{-2} yr^{-1} for the 3.2 km^2 catchment of Llyn Geirionydd. Using a combination of radiometric, magnetic and heavy metal analysis, the sequence identifies two highs of sedimentation for 1765–1830 and 1903–85, which correspond with periods of mining activity in the catchment. There is limited evidence for earlier impacts associated with deforestation or agricultural expansion, leading to the conclusion that the availability and transport of sediment in stream channels has been more important than hillslope processes (Dearing, 1992). At nearby Llyn Peris, pre-twentieth century sediment yields were low compared to more recent inputs associated with overgrazing and construction work (Dearing et al., 1981). In an attempt to relate British Holocene erosion rates from lake-based studies to models of sediment flux, Barlow and Thompson (2000) report upland lake sediment yields in the range 10–100 t km^{-2} yr^{-1}, but are unable to find field evidence for the estimated quantities of sediment in intermediate storage between slope and lake basin. The problem of interpreting sediment budgets reinforces two important points about upland sediment transfer – coupling and scale.

The delivery ratio reflects the degree of coupling between hillslopes and the channel network and the amount of this sediment that is exported through the network. Sediment stores within the catchment have different activation potentials (Richards, 1993), so that connectivity between parts of the sediment budget are a function of run-off magnitude and sediment calibre. For example, small amounts of sediment transfer from a hillslope to a colluvial store may occur several times per year, but the transfer of material out of the colluvial store into the fluvial network requires a higher-magnitude event capable of lateral incision. Preserved evidence of millennial geomorphological change results from some events that accomplished identifiable impacts to slopes or headwater tributaries but not to valley-floor sequences and vice versa. Sediment delivery can also be conceptualized as a series of kinematic waves (Richards, 1993) whereby some of the sediment yield represents rapid transfer from slope mobilization to catchment outlet, while other components represent the tail of sediment movement initiated under Late Glacial conditions. Caine and Swanson (1989) contrasted the sediment budgets of two small American catchments in the Cascades and Rockies. The former displayed an equilibrium condition where sediment production was approximately balanced by sediment export, but the latter conformed to a slow decay model of landscape development in which only 10% of the mobilized material was exported. Some British upland environments display episodic activity that couples slope erosion to channel networks, but primarily sediment transfer is a slow adjustment to postglacial conditions that is evident from the limited modification of many Late Glacial landforms after more than 10 000 years of slope processes.

Clues to upland sediment transfer preserved in depositional environments are also scale-dependent. Viewed from downstream through the cut-and-fill sequences of major river valleys, Macklin and Lewin (1993) remarked that the UK Holocene record is 'climatically driven but culturally blurred'. To some extent, impacts of land-use change on hillslope processes are more likely to be preserved in headwater fan and terrace sequences or in lake sediments of small and well-coupled catchments, whereas downstream terraces and floodplain sediments reflect changing fluvial transport regimes and the availability of sediment in channels. It can also be argued that the resolvable detail of reconstruction, and hence the ability to ascribe observed patterns to specific driving forces, is itself scale-dependent.

7.7 Conclusion

The review of geomorphological investigations in upland Britain indicates a growing interest and awareness of late Holocene process activity. The broad synchronicity of geomorphological activity between regions is alluring and there is a temptation to seek a general cause. However, the lack of precision of dating methods means that cause and effect relations are often established by temporal correlation alone, and there has been slow progress in attempting to distinguish between climate forcing and land use impacts. Furthermore, late Holocene alluvial fan activity (Ballantyne and Whittington, 1999) points to the influence of individual extreme events which are localized in nature and unlikely to be reflected in the sediment sequences of larger catchments. Large variability in process rates is a recurrent theme in upland geomorphology (Newson, 1980) and one which applies to channels (Carling, 1986) and hillslopes (Innes, 1985) alike. Response is conditioned by the distribution and magnitude of the extrinsic impact, sediment availability and calibre, and by the connectivity between slopes and channels. For the most part, long-term sediment yields in upland rivers appear to be controlled by peak discharge and the availability of sediment in the channel, but little influenced by slope processes (Dearing and Foster, 1993). It can be argued that human impact and the climatic deterioration and increased storminess associated with the Little Ice Age produced an upland landscape that was more sensitive to change (Rumsby and Macklin, 1994). Increased storminess during this period raised the probability of geomorphological change at a given location, but the probability remained sufficiently small that the occurrence of extreme events has been essentially random and independent of anthropogenic or climate control (Carling, 1986). Regionally, the last millennium has witnessed moderate geomorphological change characterized by local responses. Such a general conclusion is not without qualification. In some sensitive peatland areas such as the southern Pennines, 'the impact of human activities on the dynamic system, during the 1000 years just gone by, far from slowing down the deteriorative trends has hastened them' (Conway, 1954).

REFERENCES

Alexander, R. W., Coxon, P. and Thorn, R. H. 1986: A bog flow at Straduff Townland, County Sligo. *Proceedings of the Royal Irish Academy*, 86B, 107–19.

Archer, D. 1992: *Land of Singing Waters: Rivers and Great Floods of Northumbria*. Northumbria: Spreddon Press.

Ashmore, P., Brayshay, B. A., Edwards, K. J., Gilbertson, D. D., Grattan, J. P., Kent, M., Pratt, K. E. and Weaver, R. E. 2000: Allochthonous and autochthonous mire deposits, slope instability and palaeoenvironmental investigations in the Borve Valley, Barra, Outer Hebrides, Scotland. *The Holocene*, 10, 97–108.

Ball, D. F. and Goodier, R. 1974: Ronas Hill, Shetland: a preliminary account of its pattern features resulting from the action of frost and wind. In R. Goodier (ed.), *Natural Environment of Shetland*. The Nature Conservancy Council, 89–106.

Ballantyne C. K. 1986: Late Flandrian solifluction on the Fannich Mountains, Ross-Shire. *Scottish Journal of Geology*, 22, 395–406.

Ballantyne, C. K. 1987: The present day periglaciation of upland Britain. In J. Boardman (ed.), *Periglacial Processes and Landforms in Britain and Ireland*. Cambridge: Cambridge University Press, 113–26.

Ballantyne, C. K. 1991: Late Holocene erosion in upland Britain: climatic deterioration or human influence. *The Holocene*, 1, 81–5.

Ballantyne, C. K. 1996: Formation of miniature sorted patterns by shallow ground freezing: a field experiment. *Permafrost and Periglacial Processes*, 7, 409–24.

Ballantyne, C. K. 1998: Aeolian deposits on a Scottish mountain summit: characteristics, provenance, history and significance. *Earth Surface Processes and Landforms*, 23, 625–41.

Ballantyne, C. K. and Harris, C. 1994: *The Periglaciation of Great Britain*. Cambridge: Cambridge University Press.

Ballantyne, C. K. and Whittington, G. 1987: Niveo-aeolian sand deposits on An Teallach, Wester Ross, Scotland. *Transactions of the Royal Society of Edinburgh: Earth Sciences*, 78, 51–63.

Ballantyne, C. K. and Whittington, G. 1999: Late Holocene floodplain incision and alluvial fan formation in the central Grampian Highlands, Scotland: chronology, environment and implications. *Journal of Quaternary Science*, 14, 651–71.

Barlow, D. N. and Thompson, R. 2000: Holocene sediment erosion in Britain calculated from lake-basin studies. In I. D. L. Foster (ed.), *Tracers in Geomorphology*. Chichester: Wiley, 455–72.

Bayfield, N. G., Watson, A. and Miller, G. R. 1988: Assessing and managing the effects of recreational use on British hills. In M. B. Usher and D. B. A. Thompson (eds.), *Ecological Change in the Uplands*. Special Publication 7, British Ecological Society. Oxford: Blackwell Scientific, 399–414.

Birks, H. J. B. 1988: Long-term ecological change in the British uplands. In M. B. Usher and D. B. A. Thompson (eds.), *Ecological Change in the Uplands*. Special Publication 7, British Ecological Society. Oxford: Blackwell Scientific, 37–56.

Bower, M. M. 1960: Peat erosion in the Pennines. *Advancement of Science*, 64, 323–31.

Bower, M. M. 1961: The distribution of erosion in blanket peat bogs in the Pennines. *Transactions of the Institute of British Geographers*, 29, 17–30.

Bradshaw, R. and McGee, E. 1988: The extent and time course of mountain blanket peat erosion in Ireland. *New Phytologist*, 108, 219–24.

Brazier, V., Whittington, G. and Ballantyne, C. K. 1988: Holocene debris cone evolution in Glen Etive, western Grampian Highlands, Scotland. *Earth Surface Processes and Landforms*, 13, 525–31.

Brune, G. M. 1953 Trap efficiency of reservoirs. *Transactions of the American Society of Civil Engineers*, 109, 1080–6.

Butcher, D. P., Labadz, J., Potter, A. W. R. and White, P. 1993: Reservoir sedimentation rates in the southern Pennine region, UK. In J. McManus and R. W. Duck (eds), *Geomorphology and Sedimentology of Lakes and Reservoirs*. Chichester: Wiley, 73–92.

Caine, N. 1963: The origin of sorted stripes in the Lake District, Northern England. *Geografiska Annaler*, 45, 172–9.

Caine, N. and Swanson, F. J. 1989 Geomorphic coupling of hillslope and channel systems in two small mountain basins. *Zeitschrift für Geomorphologie*, 33, 189–203.

Carling, P. A. 1986: The Noon Hill flash floods; July 17th 1983. Hydrological and geomorphological aspects of a major formative event in an upland landscape. *Transactions of the Institute of British Geographers*, NS, 11, 105–18.

Church, M., Kellerhals, R. and Day, T. J. 1989: Regional clastic sediment yield in British Columbia. *Canadian Journal of Earth Science*, 26, 31–45.

Conway, V. M. 1954: Stratigraphy and pollen analysis of southern Pennine blanket peats. *Journal of Ecology*, 42, 117–47.

Crisp, D. T., Rawes, M. and Welch, D. 1964: A Pennine peat slide. *Geographical Journal*, 130(4), 519–24.

Curry, A. M. 2000: Holocene reworking of drift-mantled hillslopes in Glen Docherty, Northwest Highlands, Scotland. *The Holocene*, 10, 509–18.

Dearing, J. A. 1992: Sediment yields and sources in a Welsh upland catchment during the past 800 years. *Earth Surface Processes and Landforms*, 17, 1–22.

Dearing, J. A. and Foster, I. D. L. 1993: Lake sediments and geomorphological processes: some thoughts. In J. McManus and R. W. Duck (eds), *Geomorphology and Sedimentology of Lakes and Reservoirs*. Chichester: Wiley, 5–14.

Dearing, J. A., Elner, J. K. and Happey-Wood, C. M. 1981: Recent sediment flux and erosional processes in a Welsh upland lake-catchment based on magnetic susceptibility measurements. *Quaternary Research*, 16, 356–72.

Duck, R. W. and McManus, J. 1994: A long term estimate of bedload and suspended sediment yield derived from reservoir deposits. *Journal of Hydrology*, 159, 365–73.

Dunsford, H. M. 1998: The response of alluvial fans and debris cones to changes in sediment supply, upland Britain. Unpublished Ph.D. thesis, University of Durham.

Evans, R. 1989: Erosion studies in the Dark Peak. *North of England Soils Discussion Group Proceedings*, 24, 39–61.

Evans, R. 1998: The erosional impacts of grazing animals. *Progress in Physical Geography*, 22, 251–68.

Foster, I. D. L. and Lees, J. A. 1999: Changing headwater suspended sediment yields in the LOIS catchments over the last century: a palaeolimnological approach. *Hydrological Processes*, 13, 1137–53.

Foster, I. D. L., Dearing, J. A., Grew, R. and Orend, K. 1990: The lake sedimentary database: an appraisal of lake and reservoir-based studies of sediment yield. In D. E. Walling, A. Yair and S. Berkowicz (eds), *Erosion, Transport and Deposition Processes* (Proceedings of the Jerusalem Workshop). IAHS Publication No. 189. Wallingford: International Association of Hydrological Sciences, 19–43.

Gerrard, J. 1991: The status of temperate hillslopes in the Holocene. *The Holocene*, 1, 86–90.

Grove, J. M. 1972; The incidence of landslides, avalanches and floods in western Norway during the Little Ice Age. *Arctic and Alpine Research*, 4, 131–8.

Grieve, I. C., Davidson D. A. and Gordon, J. E. 1995: Nature, extent and severity of soil-erosion in upland Scotland. *Land Degradation and Rehabilitation*, 6, 41–55.

Goodier, R. and Ball, D. F. 1975: Ward Hill, Orkney: patterned ground features and their origins. In: R. Goodier (ed.), *Natural Environment of Orkney*. London: The Nature Conservancy Council, 47–56.

Harvey, A. M. 1985: The river systems of North-west England. In: R. H. Johnson (ed.), *The Geomorphology of North-west England*. Manchester: Manchester University Press, 122–42.

Harvey, A. M. 1986: Geomorphic effectiveness of a 100 year storm in the Howgill Fells, Northwest England. *Zeitschrift für Geomorphologie*, 30, 71–91.

Harvey, A. M., Oldfield, F., Baron, A. F. and Pearson, G. W. 1981: Dating of post-glacial landforms in the central Howgills. *Earth Surface Processes and Landforms*, 6, 401–12.

Harvey, A. M. and Renwick, W. H. 1987: Holocene alluvial fan and terrace formation in the Bowland Fells, northwest England. *Earth Surface Processes and Landforms*, 12, 249–57.

Howard, A. J., Macklin, M. G., Black, S. and Hudson-Edwards, K. A. 2000: Holocene river development and environmental change in Upper Wharfedale, Yorkshire Dales, England. *Journal of Quaternary Science*, 15, 239–52.

Hudson-Edwards, K. A., Macklin, M. G. and Taylor, M. P. 1999: 2000 years of sediment-borne heavy metal storage in the Yorkshire Ouse basin, NE England, UK. *Hydrological Processes*, 13, 1087–102.

Innes, J. L. 1983: Lichenometric dating of debris flow deposits in the Scottish highlands. *Earth Surface Processes and Landforms*, 8, 579–88.

Innes, J. L. 1985: Magnitude–frequency relations of debris flows in Northwest Europe. *Geografiska Annaler*, 67A (1–2), 23–32.

Innes, J. L. 1989: Rapid mass movements in upland Britain: a review with particular reference to debris flows. *Studia Geomorphologica Carpatho-Balanica*, 23, 53–67.

James, P. A. and Alexander, R. W. 1998: Soil erosion and runoff in improved pastures of the Clwydian Range, North Wales. *Journal of Agricultural Science*, 130, 473–88.

King, R. B. 1971: Vegetation destruction in the sub-alpine and alpine zones of the Cairngorm Mountains. *Scottish Geographical Magazine*, 87, 103–115.

Lawler, D. M. 1993: Needle ice processes and sediment mobilization on river banks – the River Ilston, West-Glamorgan, UK. *Journal of Hydrology*, 150, 81–114.

Legg, C. J., Maltby, E. and Proctor, M. C. F. 1992: The ecology of severe moorland fire on the North York Moors – seed distribution and seedling establishment of *Calluna vulgaris*. *Journal of Ecology*, 80, 737–752.

Lewin, J. 1980: Available and appropriate timescales in geomorphology. In R. A. Cullingford, D. A. Davidson and J. Lewin (eds), *Timescales in Geomorphology*. Chichester: Wiley, 3–10.

Luoto, M. and Seppala, M. 2000: Summit peats ('peat cakes') on the fells of Finnish Lapland: continental fragments of blanket mires? *The Holocene*, 10, 229–41.

Mackay, A. W. and Tallis, J. H. 1996: Summit-type blanket mire erosion in the Forest of Bowland, Lancashire, UK: predisposing factors and implications for conservation. *Biological Conservation*, 76, 31–44.

Macklin, M. G. 1999: Holocene river environments in prehistoric Britain: human interference and impact. In K. J. Edwards and J. P. Sadler (eds),

Holocene Environments of Prehistoric Britain. Quaternary Proceedings 7. Chichester: Wiley, 521–30.

Macklin, M. G. and Lewin, J. 1993: Holocene river alluviation in Britain. *Zeitschrift für Geomorphologie Supplementband*, 88, 109–22.

Macklin, M. G. and Rose, J. 1986: *Quaternary River Landforms and Sediments in the Northern Pennines, England: Field Guide.* British Geomorphological Research Group/Quaternary Research Association.

Macklin, M. G., Passmore, D. and Newson, M. D. 1998: Controls of short- and long-term river instability: processes and patterns in gravel-bed rivers, Tyne Basin, England. In P. C. Klingeman, R. L. Beschta, P. D. Komar and J. B. Bradley (eds.), *Gravel-bed Rivers in the Environment.* Water Resources Publications, LLC, 257–78.

Macklin, M. G., Rumsby, B. T. and Heap, T. 1992: Flood alluviation and entrenchment: Holocene valley-floor development and transformation in the British uplands. *Geological Society of America Bulletin*, 104, 631–43.

Macklin, M. G., Taylor, M. P., Hudson-Edwards, K. A. and Howard, A. J. 2000: Holocene environmental change in the Yorkshire Ouse basin and its influence on river dynamics and sediment fluxes to the coastal zone. In I. Shennan and J. Andrews (eds.), *Holocene Land–Ocean Interaction and Environmental Change Around the North Sea.* Special Publications, 166. London: Geological Society, 87–96.

Merrett, S. P. and Macklin, M. G. 1999: Historic river response to extreme flooding in the Yorkshire Dales, Northern England. In A. G. Brown and T. A. Quine (eds.), *Fluvial Processes and Environmental Change.* Chichester: Wiley, 345–60.

Newson, M. D. 1980: The geomorphological effectiveness of floods – a contribution stimulated by two recent events in mid-Wales. *Earth Surface Processes*, 5, 1–16.

Newson, M. D. and Bathurst, J. C. 1990: Sediment movement in gravel bed rivers. *Department of Geography Seminar Paper*, 59, University of Newcastle-upon-Tyne.

Pye, K. and Paine, A. D. M. 1983: Nature and source of aeolian deposits near the summit of Ben Arkle, Northwest Scotland. *Geologie en Mijnbouw*, 68, 13–18.

Ratcliffe, D. A. and Thompson, D. B. A. 1988: The British uplands: their ecological character and international significance. In M. B. Usher and D. B. A. Thompson (eds.), *Ecological Change in the Uplands.* Special Publication 7, British Ecological Society. Oxford: Blackwell Scientific, 9–36.

Reid, L. M. and Dunne, T. 1996: *Rapid Evaluation of Sediment Budgets.* Riskirchen: Catena Verlag.

Rhodes, A. N. and Stevenson, A. C. 1997: Palaeoenvironmental evidence for the importance of fire as a cause of erosion of British and Irish blanket peats. In H. J. Tallis, R. Meade and P. D. Hulme (eds.), *Blanket Mire Degradation: Causes, Consequences and Challenges.* Aberdeen: Macaulay Land Use Research Institute, 64–78.

Richards, K. S. 1993: Sediment delivery and the drainage network. In K. J.

Beven and M. J. Kirkby (eds), *Channel Network Hydrology*. Chichester: Wiley, 221–54.

Roberts, N. 1998: *The Holocene: an Environmental History*. Oxford: Blackwell.

Rowan, J. S., Goodwill, P. and Greco, M. 1995a: Temporal variability in catchment sediment yield determined from repeated bathymetric surveys: Abbeystead Reservoir, UK. *Physics and Chemistry of the Earth*, 20, 199–206.

Rowan, J. S., Barnes, S. J. A., Hetherington, S. L., Lambers, B. and Parsons, F. 1995b: Geomorphology and pollution – the environmental impacts of lead mining, Leadhills, Scotland. *Journal of Geochemical Exploration*, 52, 57–65.

Royal Commission on Environmental Pollution 1996: *Sustainable Use of Soil*. London: HMSO.

Rumsby, B. T. and Macklin, M. G. 1994: Channel and floodplain response to recent abrupt climate change: the Tyne Basin, Northern England. *Earth Surface Processes and Landforms*, 19, 499–515.

Sansom, A. L. 1996: Floods and sheep – is there a link? *Bulletin, British Ecological Society*, 27, 27–32.

Sansom, A. L. 1998: Upland vegetation management: the impacts of over-stocking. *Water Science Technology*, 39(12), 85–92.

Shimwell, D. 1974: Sheep grazing intensity in Edale 1647–1747 and its effect on blanket peat erosion. *Derbyshire Archaeological Journal*, 94, 35–40.

Stevenson, A. C., Jones, V. J. and Battarbee, R. W. 1990: The cause of peat erosion: a palaeoliminological approach. *New Phytologist*, 114, 727–35.

Sugden, D. E. 1971: The significance of periglacial activity on some Scottish mountains. *Geographical Journal*, 137(3), 388–92.

Tallis, J. H. 1994: Pool and hummock patterning in a southern Pennine blanket mire. 2. The formation and erosion of the pool system. *Journal of Ecology*, 82, 789–803.

Tallis, J. H. 1995: Climate and erosion signals in British blanket peats: the significance of *Racomitrium lanuginosum* remains. *Journal of Ecology*, 83, 1021–30.

Tallis, J. H. 1997a: Peat erosion in the Pennines: the badlands of Britain. *Biologist*, 44, 277–9.

Tallis, J. H. 1997b: The pollen record of *Empetrum nigrum* in southern Pennine peats: implications for erosion and climate change. *Journal of Ecology*, 85, 455–465.

Tallis, J. H. 1998 Growth and degradation of British and Irish blanket mires. *Environmental Review*, 6, 81–122.

Thompson, D. B. A., Galbraith, H. and Horsfield, D. 1987: Ecology and resource of Britain's mountain plateaus: land use conflicts and impacts. In M. Bell and R. G. H. Bunce (eds), *Agriculture and Conservation in the Hills and Uplands*. Swindon: ITE, NERC, 22–31.

Tipping, R. 1994: Fluvial chronology and valley floor evolution of the upper Bowmont valley, Borders Region, Scotland. *Earth Surface Processes and Landforms*, 19, 641–57.

Tipping, R. and Halliday, S. P. 1994: The age of alluvial fan deposition at a site in the southern uplands of Scotland. *Earth Surface Processes and Land-forms*, 19, 333–48.

Tomlinson, R. W. 1981: The erosion of peat in the uplands of Northern Ireland. *Irish Geography*, 14, 51–64.

Tomlinson, R. W. and Gardiner, T. 1982: Seven bog slides in the Slieve-an-Orra Hills, County Antrim. *Journal of Earth Science, Royal Dublin Society*, 5, 1–9.

Walling, D. E. 1983: The sediment delivery problem, *Journal of Hydrology*, 65, 209–37.

Walling, D. E. and Webb, B. W. 1981: Water quality. In J. Lewin (ed.), *British Rivers*. London: George Allen & Unwin, 126–69.

Warburton J. 1987: Characteristic ratios of width to depth-of-sorting for sorted stripes in the English Lake District. In J. Boardman J. (ed.), *Periglacial Processes and Landforms in Britain and Ireland*. Cambridge: Cambridge University Press, 163–71.

Warburton, J. and Caine, N. 1999: Sorted patterned ground in the English Lake District. *Permafrost and Periglacial Processes*, 10, 193–7.

Wilson, P. 1989: Nature, origin and age of Holocene aeolian sand on Muckish Mountain, Co. Donegal, Ireland. *Boreas*, 18, 159–68.

Chapter 8

Fine Particulate Sediment Transfers in Lowland Rural Environments

Ian D. L. Foster

8.1 Introduction

The amount of sediment transported by rivers will, in part, reflect spatial
and temporal differences in soil erosion, but the exact relationship
between soil erosion and sediment yield, and the factors controlling this
relationship, remain poorly understood. What is well known is that only
a proportion of the soil eroded within a catchment will find its way to
the basin outlet (Walling, 1983, 1990), since sediment can be redistributed within fields or stored at some intermediate location within the
hillslope subsystem (e.g. hedgerows and riparian buffer strips) or within
the conveyance subsystem (floodplains, wetlands, lakes and reservoirs
and the channel bed). The conveyance subsystems may provide
additional inputs (e.g. channel erosion), or improved connectivity (trac-
tor wheelings, metalled/unmetalled roads, tracks and land drains), which
may contribute to an increase in the total sediment yield of the river.
  Temporal discontinuities in sediment conveyance may introduce
additional complexity into the relationship between upstream erosion
and downstream sediment yield, because the amount of sediment
exported from a catchment may reflect the recent history of erosion
and sediment delivery rather than contemporary erosion within the
catchment (Walling, 1990). This issue has recently been addressed by
Marutani et al. (1999) in a study of the upper Waipaoa river, North
Island, New Zealand. Their analysis of sediment delivery under storm
run-off events of different magnitude and frequency suggested that two
tributaries of the Waipaoa river (Matakonekone and Oil Springs
streams) responded to extreme storms by instantly aggrading but
gradually excavated the temporarily stored sediment under subsequent
lower-magnitude events.

This chapter focuses on problems in interpreting temporal changes in sediment yield history over the last *c.*100 years largely, although not exclusively, through a review of recent UK literature. Explanations of changing sediment yield histories are briefly considered in relation to changing agricultural practice and changing weather and climate over similar timescales. Finally, an attempt is made to explore the significance of sediment storage and redistribution (sediment delivery) in the hillslope and river corridor subsystems and identify the possible contributions from the conveyance/transfer system to the total sediment yield of UK rivers.

8.2   Temporal Changes in Sediment Yield

Temporal patterns in erosion and sediment transport, particularly in the UK, are poorly documented. While sediment transport monitoring programmes have been implemented in many parts of the world, the UK has no national monitoring scheme to provide reliable estimates of sediment yield and there are few long-term databases from which to assess recent or historical trends (Walling, 1995). What data do exist have largely come from short-term research programmes over the last 30–40 years. This presents three major problems in assessing sediment yields and in identifying trends through time. First, short-term studies, often of only 2–3 years duration, may not provide reliable estimates, since differences between annual yields measured over these time periods may simply reflect short-term climatic variability. Second, different methods have been used to measure and to calculate the amount of sediment transported by rivers over this 30–40 year period, and the estimates may have unacceptably large, or unknown, errors leading to uncertainty in the reliability of the data. Third, many of the impacts that geomorphologists want to investigate occurred prior to the 1960s yet no UK data sets exist for this time period.

In a review of available UK suspended sediment data, Walling (1990) suggested that yields typically lie in the range 50–100 t km^{-2} yr^{-1} with values at the higher end of the range generally associated with upland regions. Over most of lowland Britain, sediment yields rarely exceed 50 t km^{-2} yr^{-1}, which is equivalent to an average soil erosion rate over the whole catchment of 0.5 t ha^{-1} yr^{-1}. Erosion rates reported for arable fields within the UK are often considerably higher than this and often approach, or even exceed, 50 t ha^{-1} yr^{-1} (Evans and Cook, 1986; Boardman, 1990; Foster et al., 1997).

Since the development of reliable dating methods in the 1970s, historical trends in catchment sediment yields during the twentieth century were reconstructed from studies of lake and reservoir bottom sediments, using the radionuclides Pb-210 and Cs-137 to date the depositional sequences. This approach provides sediment yield estimates averaged over timescales of c.5–50 years and, coupled with additional secondary land use and climate data, may provide information relating changes in yield to the dominant causal mechanism.

Examples of reconstructed sediment yields based on dated lake and reservoir bottom sediments are given in figure 8.1. From the examples presented here, there is evidence for an increase in sediment yields over the last 30–40 years, often by as much as 100% of the pre-1960 values. Explanations for these increases usually invoke one of two major controls; impacts of agriculture and impacts of weather, climate and climate change.

8.2.1 Agriculture

Of all factors most commonly cited as the dominant cause of increased erosion since the Second World War, agricultural practices and methods of land management are usually held responsible for increasing rates of soil erosion and sediment transport to the greatest extent. Just over a third of agricultural land in England and Wales is estimated to be at moderate to high risk from water erosion (Evans, 1990) and specific factors cited in soil erosion research include (Heathwaite et al., 1990; Boardman, 1991; Foster and Walling, 1994; Foster, 1995; Morgan, 1995; Boardman et al., 1996; Foster et al., 1996, 1997):

- the loss of soil structural stability caused by modern tillage methods and by the use of inorganic rather than organic fertilisers
- inappropriate timing and methods of cultivation (e.g. no contour ploughing, creating compacted surfaces and fine tilths)
- the removal of effective buffers to sediment and water movement, such as hedgerows
- changing the planting times of cereal crops from spring to autumn, when high-intensity rainfall is more common in the UK
- sowing row crops such as maize
- overstocking and overgrazing grasslands, which promotes greater run-off and sediment generation
- allowing stock to destroy riparian margins

The significance of land use controls has been demonstrated for small to medium-sized catchments by Naden and Cooper (1999) in an analysis of sediment concentrations in a subset of 62 sub-catchments

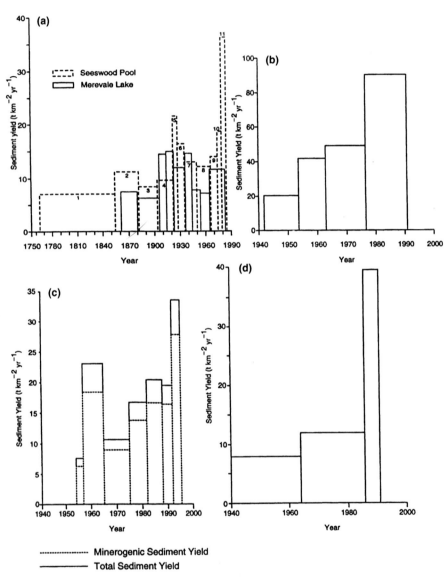

Figure 8.1 Reconstructed sediment yield histories in (a) Merevale Lake and Seeswood Pool, North Warwickshire, (b) the Old Mill Reservoir, Devon, (c) Fillingham Lake, Lincolnshire and (d) Ponsonby Tarn, Cumbria. (Based on Foster et al., 1985, 1986; Foster and Walling, 1994; Foster and Lees, 1999; Oldfield et al., 1999.)

of up to 400 km² in an area within the Yorkshire River Ouse drainage basin. A range of factors were statistically related to mean suspended sediment concentration. Development of a multiple regression model demonstrated that two factors, the percentage cropped area and the percentage urbanized area, provided the most significant control on suspended sediment concentrations at this scale and, in combination, explained 71.5% of the variance in the data.

8.2.2   Weather and climate

While the major focus on causes of erosion has been the relative significance of agricultural activities and practices, the possible impacts of changing weather patterns and more sustained shifts in climate have also been implicated. In part, climate shifts may be associated with changes in cropping practice, as seen by the widespread introduction of maize into current UK farming systems. However, highly localized erosion episodes on agricultural land associated with extreme rainfall have been well documented for the UK (Evans, 1990, 1995; Boardman, 1995, 1996; Boardman et al., 1996; Foster et al., 1997) and are often associated with high-magnitude, low-frequency rainstorms.

An important question is whether such events have become more common or whether there has been a shift in the seasonal pattern of more erosive events. Analysis of extreme daily rainfall frequencies, monthly rainfall totals and Lamb Weather Types (Foster, 1995; Wilby et al., 1997) would support the view that central England rainfall has become more seasonal, that a higher proportion of annual rainfall has been delivered in fewer more extreme daily events since the early 1960s, and that the temporal pattern in sediment yields, reconstructed from lake sediment studies, is closely related to a higher frequency of winter cyclonic conditions (for a detailed review, see Foster and Lees, in press).

While it is tempting to assume that the central problem in explaining river sediment yields requires a detailed evaluation of the relative significance of agriculture, weather and climate, in reality studies over these timescales have generally ignored the sediment delivery system and changes in connectivity which may increase or decrease the opportunity of sediment transfer to water courses.

8.3   Sediment Delivery

Despite the complexity of the sediment transfer process, the relation-ship between soil erosion and sediment yield is often expressed as a simple ratio; the sediment delivery ratio (Walling, 1983; Boardman et al., 1990). Conventional views (Walling, 1983; Milliman and Syvitski, 1992) suggest that the sediment delivery ratio decreases with an increase in catchment size, as exemplified by a decrease in specific sediment yield with increasing catchment area. It is argued that scale effects and emergent landscape properties become more important as basins become larger since: first, major storage elements, especially river floodplains and valley-fills, do not exist or are relatively insignifi-cant in small basins; second, the connectivity between hillslopes and rivers may be reduced by the presence of floodplains of increasing widths and, third, larger catchments have lower gradients, which will induce deposition. The probability that an eroded particle will be deposited increases with increased transport distance as a result of these three factors. However, two important studies have challenged these views on the sediment delivery process. First, Church and Slaymaker (1989) observed in British Columbia that specific sediment yields increased downstream in catchments of up to 3×10^4 km² due to the remobilization of Quaternary sediments stored within the valley and channel system. Second, Dedkhov and Mozzherin (1992) suggested that river systems can be characterized by either positive or negative relationships between specific sediment yield and catchment area. Positive relationships exist in rivers where channel erosion dominates (especially in forested areas). Where slope erosion constitutes the major sediment source, most of the erosion will be concentrated in headwater areas and a proportion of the mobilized sediment will be deposited during transport through the catchment, resulting in a negative relationship between catchment area and specific sediment yield.

The importance of floodplain storage in controlling the balance between erosion and specific sediment yield is exemplified by attempts to reconstruct the sediment budgets of two catchments in the South Hams region of Devon; the Old Mill reservoir (1.6 km²) and the Start catchment (10.8 km²) over a 40 year time period (Foster and Walling, 1994; Foster et al., 1996; Owens et al., 1997).

Sediment yields from the Start catchment were estimated to be between a third and a half of those of the Old Mill catchment. Based on Cs-137 measurements in catchment soils, floodplain and lake sediments, a sediment budget was reconstructed for both catchments. Field erosion rates were estimated to be $c.80$ t km⁻² yr⁻¹ and it was estimated that $c.26\%$ of eroded soil was redeposited behind hedgerows. While sediment yields entering the main Start valley ($c.60$ t km^{-2} yr^{-1}) were similar to those of the Old Mill catchment over the same time period, a significant proportion was lost to floodplain storage. Less than 30 t km^{-2} yr^{-1} (sediment delivery ratio of 38%) left the catchment. By contrast, the sediment delivery ratio of the Old Mill catchment was estimated to be $c.74\%$.

While sediment budgets of this type provide some basic grasp of the sediment delivery problem, the necessary time-averaging and spatial-lumping involved with this approach is unlikely to produce sufficient information for modelling the sediment delivery process. It also ignores the potentially large number of stores and transfers operating at different spatial and temporal scales.

Since models of erosion and sediment yield at the catchment scale need explicitly to account for the sediment delivery process, the possible transfers operating in the drainage basin in different locations and at different scales need to be quantified in order to provide a basis from which the processes might be modelled. These transfers must include all elements of the catchment, and there is a powerful argument to suggest that empirically based studies of erosion and sediment yield should move towards multidisciplinary, multi-scale (spatial and temporal) approaches that focus more on the linkages and less on the individual processes.

8.4   Sediment Transfers

8.4.1   Hillslope/river linkages

Direct links between the hillslope and river may occur as a result of overland flow, rill and gully erosion, although even at the field scale there is substantial evidence to suggest that not all eroded sediment reaches a drainage ditch or river, or even the adjacent field. Soil redistribution and sediment delivery at the field scale using the Cs-137 technique has been estimated at a number of UK locations by Quine and Walling (1991) (table 8.1). Delivery ratios for six soil types on individual fields range from 30% to 85%. Since hedgerows, adjacent fields and riparian buffer strips will control the amount of sediment that leaves the field and enters a river, the transfer of sediment may be controlled as much by the position of the gateway to the field, or the proximity of an open road drain, as by the erosional processes operating within the field. In consequence, fields located some distance away from a major water course may have a high connectivity if, for example,

Table 8.1 Sediment delivery data for arable fields on various soil types in southern Britain (after Quine and Walling, 1991)

Soil type/site	Gross erosion (t ha^{-1} yr^{-1})	Net erosion (t ha^{-1} yr^{-1})	Sediment delivery ratio (%)	Retention (%)
Brown Sand				
Hole, Norfolk	6.3	3.0	48	52
Dalicott, Shropshire	10.2	6.5	64	36
Rufford Forest, Nottinghamshire	12.2	10.5	85	15
Brown Rendzina				
Manor House, Norfolk	6.3	2.4	39	61
Lewes, Sussex	4.3	1.4	33	67
Brown calcareous earth				
West Street, Kent	7.7	4.3	55	45
Brown Earth				
Mountfield, Somerset	4.6	2.2	49	51
Yendacott, Devon	5.3	1.9	36	64
Argillic Brown Earth				
Wooton, Herefordshire	6.4	2.8	43	57
Fishpool, Gwent	5.1	1.9	37	63
Calcareous pelosol				
Higher, Dorset	5.2	3.1	59	41
Brook End, Bedfordshire	3.6	1.2	33	67
Keysoe Park, Bedfordshire	2.2	0.6	30	70

the gateway is at the lowest topographic point within the field boundary and there is a connection (a track, road or roadside drain) leading directly to a water course. To date, few studies of connectivity at the field/small catchment scale are known to have been undertaken (Harden, 1992a). Single scale studies, using either rainfall simulators or field observations, have examined the relationship between land surface conditions, the timing and volume of run-off, and the amount of sediment that might be generated (Heathwaite et al., 1990; Harden, 1992b). However, the connectivity between the point of sediment production and the nearest stream channel is usually ignored. Some useful evidence for the significance of roads is provided by Froehlich (1995) for the Homerka research catchment in the Polish Carpathians, which has shown that unmetalled roads act both as a source and as a conduit for fine particulate sediment directly entering the river network.

It is generally assumed that the connection between fields and water courses occurs via surface pathways. In the UK, however, 50% of agricultural land has been drained (often using clay or plastic pipes installed at depths of between 70 and 120 cm below the ground surface). These have been used for a number of reasons, including removal of excess water in spring, a reduction in groundwater levels and the removal of surface water (Robinson and Armstrong, 1988; Briggs and Courtney, 1991; Robinson and Rycroft, 1999). The most significant increase in land drainage in the UK occurred between about 1968 and 1973 as a result of the availability of government subsidies (Robinson and Armstrong, 1988). Despite extensive research into land drains over the last 150 years, particularly into their effects on soil moisture, groundwater levels, river hydrographs and dissolved nutrient losses, no systematic research has been undertaken on their role as sediment conduits. Recent interest in tile drains has been prompted by observations that drainage water often contains high levels of undesirable nutrients and contaminants (Dils and Heathwaite, 1996; Hardy, 1997) and that these have been observed to arrive at the base of the soil column or in tile drain outflow more rapidly than that predicted by equations considering matrix flow only (Beven and Germann, 1982; McCoy et al., 1994). However, the exact significance of drains as a conduit for fine particulate sediment is not known. Chapman et al. (2001) have shown that sediment concentrations from a single land drain at the ADAS Rosemaund experimental catchment in Herefordshire can exceed 10 g l^{-1} and, over the period 1997–8, sediment yields were estimated to be $c.159$ t km^{-2} yr^{-1} with an associated particulate phosphorus loss of $c..57$ kg ha^{-1}. The median particle size of tile drain sediments was less than 10 μm and preliminary attempts to fingerprint sediment sources from an analysis of the mineral magnetic, radionuclide and geochemical signatures of the transported sediments suggested that they were dominated by topsoil, a result confirmed in other recent tracer studies on tile drain sediments (Hardy et al., 2000). Several published references in mainland Europe and North America have identified the significance of land drains as a source of diffuse particulate phosphorus (Bottcher et al., 1981; Bengtson et al., 1992; Grant et al., 1996; Øygarden et al., 1997; Laubel et al., 1999) and a number of recent studies have identified a range of UK soil types in which sediment has been observed in tile drain flow (table 8.2). The majority of at-risk soils lie in a wide belt across the Midlands of England and the Welsh borders (Hardy, 1997; Hardy et al., 2000).

Table 8.2 The UK soil series from which sediment has been observed in subsurface land drains

Soil series (subclass)	Coverage (%)	Soil type	Reference
Melford (5.71)	0.07	Clay loam over clayey	Hardy et al. (2000)
Ludford (5.71)	0.25	Fine loamy head	Hardy (1997)
Bromyard (5.71)	0.53	Stoneless silty clay loam	Williams et al. (1996), Chapman et al. (2001)
Denchworth (7.12)	1.79	Fine loamy over clayey	Jones et al. (1995)
Dunkeswick (7.11)	1.74	Fine loamy over clayey	Brown et al. (1995)
Worcester (4.31)	0.48	Clayey	Sanders (1997)
Whimple (5.72)	0.80	Fine loamy	Sanders (1997)
Efford (5.71)	0.08	Fine loamy	Parsons (pers. comm.)
Bishampton (5.72)	0.14	Fine loamy	Parsons (pers. comm.)
Parkgate (8.41)	0.09	Stoneless silty gley	Parsons (pers. comm.)
Salop (7.11)	0.80	Stony clay loam	Dils and Heathwaite (1999)
Hodnet (5.72)	0.19	Stony clay loam	Dils and Heathwaite (1999)
Compton (8.13)	0.16	Clayey	Chapman et al. (2001)
Middleton (5.72)	0.21	Fine Silty	Chapman et al. (2001)
Hallsworth	0.73	Clayey	Chapman et al. (2001)

8.4.2 The river corridor

A number of elements serve to offer storage potential for fine sediments moving through the drainage network, but again may function at different temporal and spatial scales. These include lakes and reservoirs, wetlands, floodplains and the channel bed. Additional erosional processes, especially river bank erosion, may also be important within the channel network.

With only limited exceptions (e.g. Llangorse Lake, the Cheshire and Shropshire Meres, the Norfolk Broads, freshwater barrier beach lagoons such as Slapton Ley and Loe Pool, and periodically inundated floodplain lakes), natural lakes are not common in UK lowland environments. Artificial water bodies, however, have been constructed in the UK since at least medieval times. These include fish ponds, mill ponds, canal feeder and water supply reservoirs, ornamental ponds and urban detention basins. They exist in various locations throughout the drainage basin and frequently provide suitable sites for palaeoenvironmental reconstruction. Although sites of this type have been used by many researchers to reconstruct drainage basin sediment yield histories, few studies have attempted to calculate the impacts of these sinks on sediment transfers through catchments. Notable exceptions include the work of Meade and Parker (1985), who provide a number of case studies demonstrating decreases in sediment yield following reservoir construction in the USA. The Rio Grande, for example, currently discharges less than 1×10^6 tonnes of sediment to the Gulf of Mexico whereas before 1940 the figure was $c.40 \times 10^6$ tonnes. The most recent estimate suggests that there are at least 39 000 dams world-wide greater than 15 m high (Takeuchi et al., 1998) and several million smaller reservoirs probably also exist world-wide, producing more localized impacts on downstream sediment yields.

Summary data of table 8.3 suggest that UK lakes and reservoirs act as important sediment sinks and that accumulation rates can reach several centimetres per year. Furthermore, the modelling exercise undertaken by Naden and Cooper (1999) reported earlier omitted heavily reservoired sub-catchments from their analysis, since the presence of reservoirs had a significant impact on downstream suspended sediment concentrations and could not be modelled using simple land use controls.

Conveyance losses through the channel network occur through both over-bank and bed sediment storage (Walling and Amos, 1999, Walling et al., 1999). The floodplains of most lowland rivers of Britain are characterized by extensive deposits of fine sediment resulting from the

Table 8.3 Sediment accumulation rates for selected lowland UK lakes and reservoirs, estimated using dated sediment cores

Site	County	Mean sedimentation rate ($cm\ yr^{-1}$)	Source
Slapton Lower Ley	Devon	0.8–1.2	Heathwaite (1993)
Old Mill Reservoir	Devon	1.7	Foster and Walling (1994)
Stourton Lake	Wiltshire	1.7	He et al. (1996)
Chard Reservoir	Somerset	0.2–2.0	He et al. (1996)
Wadhurst Park Lake	West Sussex	1.4–1.7	He et al. (1996)
Merevale Lake	Warwickshire	0.6	Foster et al. (1985)
Seeswood Pool	Warwickshire	1.3	Foster et al. (1986)
Elleron Lake	North Yorkshire	1.1	Foster and Lees (1999)
Boltby Reservoir	North Yorkshire	1.8	Foster and Lees (1999)
Newburgh Priory Pond	North Yorkshire	0.6	Foster and Lees (1999)
Silsden Reservoir	West Yorkshire	0.7	Foster and Lees (1999)
March Ghyll Reservoir	West Yorkshire	0.6	Foster and Lees (1999)
Fillingham Lake	Lincolnshire	0.6	Foster and Lees (1999)
Barnes Loch	Borders	0.4	Foster and Lees (1999)
Yetholm Loch	Borders	0.7	Foster and Lees (1999)
Fontburn Reservoir	Northumbria	1.0	Foster and Lees (1999)
Ponsonby Tarn	Cumbria	0.7	Oldfield et al. (1999)

deposition of suspended sediment during overbank flow. Fallout radionuclides offer significant potential for estimating rates of deposition
over decades to centuries, and more recently have been used to estimate
the relative significance of floodplain storage in relation to the sediment
yields of rivers (He and Walling, 1996a,b). Estimates for the rivers
Ouse and Tweed (table 8.4) suggest that conveyance losses by overbank
sedimentation range from 31% to to 49% of the annual suspended
sediment yield.

Assessing the relative significance of storage within the river channel
is more problematic, although a number of authors have attempted to
develop methods that will provide estimates of fine sediment accumulation in rivers whose bed sediments comprise coarse river gravels
(Lambert and Walling, 1988; Russell et al., 1998, Walling and Amos,
1999). Some recent estimates of the contribution of channel storage to
the possible conveyance loss of rivers is given in table 8.5. The data
suggest that the contribution of stored fine sediment in river gravels is
considerably less important than floodplain storage, although estimates
from the Yorkshire Ouse suggest that as much as 10% of the annual
suspended sediment load could be stored in the river bed.

The channel network can also make a significant contribution to the
river system as well as offering important storage opportunities. Table
8.6 gives estimates of the contribution of bank erosion from the River
Culm in Devon, where c.19% of the sediment yield could be attributed
to bank erosion. Combined with floodplain sedimentation data, the
significance of conveyance losses and channel erosion relative to the
total sediment yield was also estimated for the upper and lower
catchments (figure 8.2). More recently, estimates of the relative significance of channel bank and other contributions to the sediment yield of
rivers have been made using sediment fingerprinting techniques.
Recently published data for a number of Yorkshire rivers suggest that
as much as 37% of fine sediment load is derived from channel bank
sources (table 8.7). However, more detailed research on the spatial
distribution of river bank erosion in the Swale–Ouse system of northern
England has shown that the piedmont zone of the river basins was
especially active (Lawler et al., 1999). Consequently, the relative
contribution of bank erosion to the sediment yield of the river system
is likely to vary with position in the drainage network.

8.5 Discussion and Conclusion

A synthesis of the retention mechanisms, contributions to conveyance
losses and transfer processes outlined above is given in table 8.8. While

Table 8.4 Estimates of the total annual storage of sediment on floodplains and mean annual suspended sediment yields for 1995 and 1996. (a) Based on data in Walling et al. (1999a); (b) other published UK studies

(a)

Basin/river	Floodplain storage (t yr^{-1})	Suspended sediment load (t yr^{-1})	Total delivered to channel system (t yr^{-1})	Conveyance loss in main channel (%)
Ouse				
Swale	19 214	42 352	61 566	31
Nidd	7 573	7 719	15 292	50
Ure[a]	15 125	28 887	44 012	34
Ouse[a]	18,733			
Ouse[b]	49 041	75 111	124 152	39
Ouse[c]	60 645			
Wharfe	10 325	10 816	21 141	49
Tweed				
Tweed	43 920[d]	66 012[e]	109 932	40

(b)

	Floodplain storage (t yr^{-1})	Conveyance loss in main channel (%)	Source
Severn	–	23	Walling and Quine (1993)
Start	340	38	Foster et al. (1996), Owens et al. (1997)
Culm	2 500	38	Lambert and Walling (1987)

[a] 'Ure' refers to the River Ure at its confluence with the River Swale; 'Ouse' refers to the River Ure/Ouse below this point to the tidal limit.
[b] Total to Ouse Gauging Station.
[c] Total to tidal limit.
[d] Represents the sum of floodplain storage for Ettrick Water and the rivers Teviot and Tweed.
[e] Based on Harmonised Monitoring Programme Data for 1995 and 1996.

no attempt has been made to quantify the magnitude of conveyance losses, their importance has been identified on a relative scale of low to high. It is important to recognize that many of the changes over the last 100 years in British catchments need to be set in the context of longer-

Table 8.5 Estimates of channel-bed storage in UK rivers[a]

River	Sediment storage (g cm⁻²)	Sediment storage (t km⁻¹)	Conveyance loss (%)	Source
Exe[b]	0.040	5.6	<2	Lambert and Walling (1988)
Tweed[c]	0.064	16.2	3	Owens et al. (1998)
Teviot[c]	0.112	24.6	8	Owens et al. (1998)
Ettrick[c]	0.057	13.1	–	Owens et al. (1998)
Swale[d]	0.155	41.3	–	Walling et al. (1998)
Nidd[d]	0.133	22.6	–	Walling et al. (1998)
Ouse[d]	0.200	97.7	10	Walling et al. (19980
Wharfe[d]	0.124	20.3	9	Walling et al. (1998)
Severn	–	–	1.8	Walling and Quine (1993)

[a] Based on main channel length multiplied by channel width.
[b] Devon (based on a 35 km long river reach).
[c] Scotland.
[d] Yorkshire.

Table 8.6 The estimated contribution from bank erosion to the suspended sediment output of the River Culm at Rewe, 1980–1 (based on data in Ashbridge, 1995)

Particle size	Net amount from banks (t)	Total catchment output (t)	Relative contribution (%), with upper and lower 95% confidence limits		
			Min.	Mean.	Max.
Silt	758	3370	14	22	35
Clay	681	4290	10	16	25
Silt + clay	1439	7660	13	19	30

term environmental change. Major changes in both land use and climate appear to have increased rates of soil erosion and sediment transport over the last 30–40 years, and the body of evidence currently available suggests that floodplains and reservoirs are currently acting as major sediment stores in the river corridor, while hedgerows are acting as important stores in the hillslope system. However, throughout the Holocene, periods of aggradation and degradation are frequently recorded in alluvial and colluvial records (see, e.g., Foster et al., 2000)

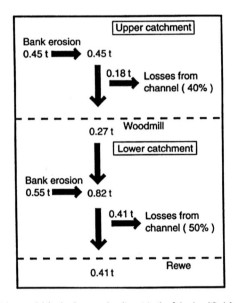

Figure 8.2   A sediment loss model for bank-sourced sediment in the Culm (modified from Ashbridge, 1995).

Table 8.7   Load-weighted mean contributions of each source type to the suspended sediment samples collected from downstream sampling sites on the rivers Swale, Ure, Nidd, Ouse and Wharfe (November 1994–February 1997) (based on Walling et al., 1999b)

		Source contributions (%)			
River	Number of samples	Woodland topsoil	Uncultivated topsoil	Cultivated topsoil	Channel-bank material
Swale	19	0	41.8	30.0	28.2
Ure	14	0.7	45.1	17.0	37.2
Nidd	14	6.9	75.2	2.8	15.1
Ouse	30	0	24.6	38.1	37.3
Wharfe	7	4.4	69.5	3.6	22.5

and probably relate to changes in the availability of sediment and to changes in river discharge over these timescales (Starkel, 1983).

To date, there has been no single study that has attempted to assess the relative significance of all of the retention mechanisms and sediment transfers discussed above in order empirically to determine the timescales over which sediment might be retained in each of the major

Table 8.8 Fine particulate sediment transfers in lowland UK drainage basins

Retention/release site	Retention mechanism	Contribution to conveyance loss	Transfer process to channel
Hillslopes			
Within field	Redistribution/ hedgerow retention; riparian buffer	Medium/high	Overland flow rills/gullies/ gateways/tracks/ roads/road drains/ subsurface land drains
River corridor			
Corridor storage	Ponds/lakes		
	Wetlands	High	Not known
	Floodplains	High	Lateral erosion
River channels	Surface accumulation; retained in gravel matrix	Low/seasonal	High-magnitude river discharge/ bed remobilization

storage elements or to determine the relative significance of individual sediment transfers in a medium-to-large river basin. Until we begin to obtain such information, it seems likely that attempts to model erosion and sediment yield will remain problematic at the catchment scale. Furthermore, any attempt to develop nutrient and contaminant transport models in situations in which the contaminant of interest has a significant particulate phase will also require this information. Despite major advances in understanding many of the factors controlling sediment yields over the last decade, there still remain a number of important unanswered questions.

REFERENCES

Ashbridge, D. 1995: Processes of river bank erosion and their contribution to the suspended sediment load of the River Culm, Devon. In I. D. L. Foster, A. M. Gurnell and B. W. Webb (eds), *Sediment and Water Quality in River Catchments*. Chichester: Wiley, 229–45.

Bengtson, R. L., Carter, C. E. and Fouss, J. L. 1992: A decade of subsurface drainage environmental research in southern Louisiana. In *Drainage and*

Water Table Control. Proceedings of the 6th International Drainage Symposium, 13–15 September, American Society of Agricultural Engineers.

Beven, K. and Germann P. 1982: Macropores and water flow in soils. *Water Resources Research,* 18, 1311–25.

Boardman, J. 1990: Soil erosion on the South Downs, a review. In J. Boardman, I. D. L. Foster and J. A. Dearing (eds), *Soil Erosion on Agricultural Land.* Chichester: Wiley, 87–105.

Boardman, J. 1991: Land use, rainfall and erosion risk on the South Downs. *Soil Use and Management,* 7, 34–8.

Boardman, J. 1995: Damage to property by runoff from agricultural land, South Downs, southern England, 1976–1993. *Geographical Journal,* 161, 177–91.

Boardman, J. 1996: Soil erosion by water: problems and prospects for research. In M. G. Anderson and Brooks, S. M. (eds), *Advances in Hillslope Processes,* Vol. 1. Chichester: Wiley, 489–505.

Boardman, J., Dearing, J. A. and Foster, I. D. L. 1990: Soil erosion studies: some assessments. In: Boardman, J., Foster, I. D. L. and Dearing, J. A. (eds) *Soil Erosion on Agricultural Land.* Chichester: Wiley, 659–72.

Boardman, J., Burt, T. P., Evans, R., Slattery, M. C. and Shuttleworth, H. 1996: Soil erosion and flooding as a result of a summer thunderstorm in Oxfordshire and Berkshire, May 1993. *Applied Geography,* 16, 21–34.

Bottcher, A. B., Monke, E. J. and Huggins, L. F. 1981: Nutrient and sediment loadings from a subsurface drainage system. *Transactions of the American Society of Agricultural Engineers,* 24, 1221–6.

Briggs, D. and Courtney, F. 1991: *Agriculture and Environment,* 2nd edn. London, Longman.

Brown, C. D., Hodgkinson, R. A., Rose, D. A., Syers, J. K. and Wilcockson, S. J. 1995: Movement of pesticides to surface waters from a heavy clay soil. *Pesticide Science,* 43, 131–40.

Chapman, A. S., Foster, I. D. L., Lees, J. A., Hodgkinson, R. A. and Jackson, R. M., 2001: Particulate phosphorus transport by sub-surface drainage from agricultural land in the UK. Environmental significance at the catchment and national scale. *Science of the Total Environment,* 266, 95–102.

Church, M. and Slaymaker, H. O. 1989: Disequilibrium of Holocene sediment yields in glaciated British Columbia. *Nature,* 337, 452–4.

Dedkhov, A. P. and Mozzherin, V. I. 1992: Erosion and sediment yield in mountain regions of the world. In *Erosion, Debris Flows and Environment in Mountain Regions (Proceedings of the Chengdu Symposium).* IAHS Publication No. 209, 29–36.

Dils, R. M. and Heathwaite, A. L. 1996: Phosphorus fractionation in hillslope hydrological pathways contributing to agricultural runoff. In M. G. Anderson and S. M. Brooks (eds), *Advances in Hillslope Processes,* Vol. 1. Chichester: Wiley, 229–51.

Evans, R. 1990: Soils at risk of accelerated erosion in England and Wales. *Soil Use and Management,* 6, 125–31.

Evans, R. 1995: Assessing costs to farmers both cumulative and in terms of risk management; downstream costs off-farm. In *Soils, Land Use and Sustainable Development*. Proceedings of the Save our Soils Conference, British Society of Soil Science, Farmers Link, April.

Evans, R. and Cook, S. 1986: Soil erosion in Britain. *SEESOIL*, 3, 28–59.

Foster, I. D. L. 1995: Lake and reservoir bottom sediments as a source of soil erosion and sediment transport data in the UK. In I. D. L. Foster A. M., Gurnell and B. W. Webb (eds), *Sediment and Water Quality in River Catchments*. Chichester: Wiley, 265–83.

Foster, I. D. L. and Lees, J. A. 1999: Changing headwater suspended sediment yields in the LOIS catchments over the last century: a palaeolimnological approach. *Hydrological Processes*, 13, 1137–53.

Foster, I. D. L. and Lees, J. A., in press: Evidence for past erosional events from lake sediments. In J. Boardman and D. Favis-Mortlock (eds), *Climate Change and Soil Erosion*. London: Imperial College Press.

Foster, I. D. L. and Walling, D. E. 1994: Using reservoir deposits to reconstruct changing sediment yields and sources in the catchment of the Old Mill reservoir, South Devon, UK over the past 50 years. *Hydrological Sciences Journal*, 39, 347–68.

Foster, I. D. L., Dearing J. A. and Appleby, P. G. 1986: Historical trends in catchment sediment yields: a case study in reconstruction from lake sediment records in Warwickshire, UK. *Hydrological Sciences Journal*, 31, 427–43.

Foster, I. D. L., Harrison, S. and Clark, D. 1997. Extreme event soil erosion on agricultural land in the UK: an act of God or agricultural mismanagement? *Geography*, 82, 231–9.

Foster, I. D. L., Mighall, T. M., Wotton, C., Owens, P. N. and Walling, D. E. 2000: Evidence for medieval soil erosion in the South Hams region of Devon, UK. *The Holocene*, 10, 255–65.

Foster, I. D. L., Dearing, J. A., Simpson, A., Carter, A. D. and Appleby, P. G. 1985: Lake catchment based studies of erosion and denudation in the Merevale Catchment, Warwickshire, UK. *Earth Surface Processes and Landforms*, 10, 45–68.

Foster, I. D. L., Walling, D. E. and Owens, P. N. 1996: Sediment yields and sediment delivery processes in the catchments of Slapton Lower Ley, South Devon, UK. *Field Studies*, 8, 629–61.

Froehlich, W. 1995: Sediment dynamics in the Polish Flysch Carpathians. In I. D. L. Foster, A. M. Gurnell and B. W. Webb (eds), *Sediment and Water Quality in River Catchments*. Chichester: Wiley, 453–61.

Grant, R., Laubel, A., Kronvang, B., Andersen, H. E., Svendsen, L. M., and Fuglsang, A. 1996: Loss of dissolved and particulate phosphorus from arable catchments by sub-surface drainage. *Water Research*, 30, 2633–42.

Harden, C. P. 1992a: A new look at soil erosion processes in highland Ecuador. In *Erosion, Debris Flows and Environment in Mountain Regions (Proceedings of the Chengdu Symposium)*. IAHS Publication No. 209, 77–85.

Harden, C. P. 1992b: Incorporating roads and footpaths in watershed-scale hydrologic and soil erosion models. *Physical Geography*, 13, 368–85.

Hardy, I. A. J. 1997: Water quality from contrasting drained clay soils: the relative importance of sorbed and aqueous phase transport mechanisms. Unpublished Ph.D. thesis, Cranfield University.

Hardy, I. A. J., Carter, A. D., Leeds-Harrison, P. B., Foster, I. D. L. and Sanders, R. M. 2000: The origin of sediment in field drainage water. In I. D. L. Foster (ed.), *Tracers in Geomorphology*. Chichester: Wiley, 241–57.

He, Q. and Walling, D. E. 1996a: Rates of overbank sedimentation on the floodplains of British lowland rivers documented using fallout caesium-137. *Geografiska Annaler*, 78A, 223–34.

He, Q. and Walling, D. E. 1996b: Use of fallout Pb-210 measurements to investigate longer term rates and patterns of overbank sediment deposition on the floodplains of lowland rivers. *Earth Surface Processes and Landforms*, 21, 141–54.

He, Q., Walling, D. E. and Owens, P. N. 1996: Interpreting the caesium-137 profiles observed in several small lakes and reservoirs in southern England. *Chemical Geology*, 129, 115–31.

Heathwaite, A. L. 1993: Lake sedimentation. In T. P. Burt (ed.), *A Field Guide to the Geomorphology of the Slapton Region*. Occasional Publication of the Field Studies Council, 27, 31–41.

Heathwaite, A. L., Burt, T. P. and Trudgill, S. T. 1990: Land use controls on sediment production in a lowland catchment, south-west England. In J. Boardman, I. D. L. Foster and J. A. Dearing (eds), *Soil Erosion on Agricultural Land*. Chichester: Wiley, 69–86.

Jones, R. L., Harris, G. L., Catt, J. A., Bromilow, R. H., Mason, D. J. and Arnold, D. J. 1995: Management practices for reducing movement of pesticides to surface water in cracking clay soils. *Proceeding of the Brighton Crop Protection Conference – Weeds*, 1, 489–98.

Lambert, C. P. and Walling, D. E. 1987: Floodplain sedimentation: a preliminary investigation of contemporary deposition within the lower reaches of the River Culm, Devon, UK. *Geografiska Annaler*, 69A, 393–404.

Lambert, C. P. and Walling, D. E. 1988: Measurement of channel storage of suspended sediment in a gravel-bed river. *Catena*, 15, 65–80.

Laubel, A. Jacobsen, O. H., Kronvang, B., Grant, R. and Anderson, H. E. 1999: Subsurface drainage loss of particles and phosphorus from field plot experiments and a tile drained catchment. *Journal of Environmental Quality*, 28, 576–84.

Lawler, D. M., Grove, J. R., Couperthwaite, J. S. and Leeks, G. J. L. 1999: Downstream change in river bank erosion rates in the Swale–Ouse system, northern England. *Hydrological Processes*, 13, 977–92.

Marutani, T., Kasai, M., Reid, L. M. and Trustram, A. 1999: Influence of storm related sediment storage on the sediment delivery from tributary catchments in the upper Waipaoa River, New Zealand. *Earth Surface Landforms and Processes*, 24, 881–96.

McCoy, E. L., Boast, C. W., Stehouwer, R. C. and Kladivko, E. J. 1994: Macropore hydraulics: taking a sledgehammer to classical theory. In R. Lal and B. A. Stewart (eds), *Soil Processes and Quality*. Boca Raton: Lewis, 303–48.

Meade, R. H. and Parker, R. S. 1985: Sediment in rivers in the United States. *US Geological Survey Water Supply Paper* 2275, 49–60.

Milliman, J. D. and Syvitski, J. P. M. 1992: Geomorphic/tectonic control of sediment discharge to the oceans: the importance of small mountainous rivers. *Journal of Geology*, 100, 325–44.

Morgan, R. P. C. 1995: *Soil Erosion and Conservation*, 2nd edn. London: Longman.

Naden, P. and Cooper, D. M. 1999: Development of a sediment delivery model for application in large river basins. *Hydrological Processes*, 13, 1011–34.

Oldfield, F., Appleby, P. G. and Van der Post, K. D. 1999: Problems of core correlation, sediment source ascription, and yield estimation in Ponsonby Tarn, West Cumbria, UK. *Earth Surface Processes and Landforms*, 24(11), 975–92.

Owens, P. N., Walling, D. E. and Leeks, G. J. L. 1998: Deposition and storage of fine-grained sediment within the main channel system of the River Tweed, Scotland. *Earth Surface Processes and Landforms*, 24, 1061–76.

Owens, P. N., Walling, D. E., He, Q., Shanahan, J. and Foster, I. D. L. 1997: The use of caesium-137 measurements to establish a sediment budget for the Start catchment, Devon, UK. *Hydrological Sciences Journal*, 42, 405–23.

Øygarden, L. Kvaerner, J. and Jenssen, P. D. 1997: Soil erosion via preferential flow to drainage systems in clay soils. *Geoderma*, 76, 65–86.

Quine, T. A. and Walling, D. E. 1991: Rates of soil erosion on arable fields in Britain: quantitative data from caesium-137 measurements. *Soil Use and Management*, 7, 169–76.

Robinson, M. and Armstrong, A. C. 1988: The extent of agricultural field drainage in England and Wales 1971–80. *Transactions of the Institute of British Geographers*, 13, 19–28.

Robinson, M. and Rycroft, D. W. 1999: The impact of drainage on stream-flow. *Agricultural Drainage, Agronomy Monograph*, 38, 767–800.

Russell, M. A., Walling, D. E., Webb, B. W. and Bearne, R. 1998: The composition of nutrient fluxes from contrasting UK river basins. *Hydrological Processes*, 12, 1461–82.

Sanders, R. M. 1997: The characterisation of drainflow sediments from agricultural soils using magnetic, radionuclide and geochemical techniques. Unpublished M.Sc. thesis, Coventry University.

Starkel, L. 1983: The reflection of hydrologic changes in the fluvial environment of the temperate zone during the last 15 000 years. In K. J. Gregory (ed.), *Background to Palaeohydrology*. Chichester: Wiley, 213–35.

Takeuchi, K., Hamlin, M., Kundzewicz, Z. W., Rosbjerg, D. and Simonovic, S. 1998: *Sustainable Reservoir Development and Management*. IAHS Publication No. 251, 190 pp.

Walling, D. E. 1983: The sediment delivery problem. *Journal of Hydrology*, 65, 209–37.

Walling, D. E. 1990: Linking the field to the river. In J. Boardman, I. D. L. Foster and J. A. Dearing (eds), *Soil Erosion on Agricultural land*, Chichester: Wiley, 129–52.

Walling, D. E. 1995: Suspended sediment yields in a changing environment. In A. M. Gurnell and G. E. Petts (eds), *Changing River Channels*. Chichester: Wiley, 149–76.

Walling, D. E. and Amos, C. M. 1999: Source, storage and mobilisation of fine sediment in a chalkstream system. *Hydrological Processes*, 13, 323–40.

Walling, D. E. and Quine, T. A. 1993: Using Chernobyl-derived fallout radionuclides to investigate the role of downstream conveyance losses in the suspended sediment budget of the River Severn, United Kingdom. *Physical Geography*, 14, 239–53.

Walling, D. E., Owens, P. N. and Leeks, G. J. L. 1998: The role of channel and floodplain storage in the suspended sediment budget of the River Ouse, Yorkshire, UK. *Geomorphology*, 22, 225–42.

Walling, D. E., Owens, P. N. and Leeks, G. J. L. 1999a: Rates of contemporary overbank sedimentation and sediment storage on floodplains of the main channel systems of the Yorkshire Ouse and River Tweed, UK. *Hydrological Processes*, 13, 993–1009.

Walling, D. E., Owens, P. N. and Leeks, G. J. L. 1999b: Fingerprinting suspended sediment sources in the catchment of the river Ouse, Yorkshire. *Hydrological Processes*, 13, 955–75.

Wilby, R. L., Dalgleish, H. Y. and Foster, I. D. L. 1997: The impact of weather patterns on historic and contemporary catchment sediment yields. *Earth Surface Processes and Landforms*, 22, 353–63.

Williams, R. J., Brooke, D. N., Clare, R. W., Matthiessen, P. and Mitchel, R. D. J. 1996: Rosemaund pesticide transport study 1987–1993. *Report No. 129*, Institute of Hydrology, Wallingford.

Chapter 9

Living with Natural Hazards:
the Costs and Management Framework

E. Mark Lee

9.1   Introduction

Geomorphological processes operating on hillslopes, within river-channel networks and at the coast present a range of management issues that influence the effective use of land in Britain. These processes shape the landscape – forming, for example, the coastal cliffs and broad meandering rivers that are part of our natural heritage. They can create and sustain valued habitats and maintain important recreational beaches or sand dunes. The processes only become hazards or problems when society encroaches into these dynamic environments either for housing or development. This chapter provides an indication of the nature and scale of the geomorphological hazards that have been a feature of the British landscape over the last millennium. Such events include:

*   slope erosion and mudfloods on hillslopes
*   dust blows on agricultural land
*   flash floods in upland areas
*   bank erosion, sedimentation and channel instability on rivers
*   floods on lowland rivers
*   sedimentation in river channels and estuaries
*   floods in low-lying coastal areas
*   erosion of coastal cliffs
*   wind-blown sand in coastal dunes

In illustrating the significance of these processes to land use and development, emphasis will be placed on damaging events over the last 200 years or so – the period covered by national and local newspaper records. However, reference will also be made to notable earlier events. As these events have imposed very high costs through loss of life and damage to property, so an administrative framework has evolved to manage the reduction of risks to both individuals and society as a whole. The origins and development of this framework will be described, drawing attention to the increasing state intervention in the management of particular problems, notably flooding, coastal erosion and landsliding.

9.2   The Historical Record

The historical archive of damaging events is very rich, ranging from early public records and monastic chronicles to recent newspaper articles, reports of flood events and research observations. By their nature these archives tend to be biased towards the unusual, or towards significant events or disasters (Brunsden et al., 1995). Thus damaging floods, landslides and coastal erosion tend to be recorded, whereas less dramatic processes (e.g. hillslope erosion and channel sedimentation) are often ignored by most sources or simply treated as a routine, and un-newsworthy, occurrence (e.g. the regular dredging of a navigation channel). In addition to this impact magnitude filter, historical records tend to be concentrated on events that have occurred in and around built-up areas, thus creating a spatial bias. Both filters may change over time, as society becomes more sensitive to smaller events and as the built-up area spreads.

Recent research for the Department of Environment, Transport and the Regions (DETR) has led to the compilation of an archive of significant erosion and flooding events for Great Britain (Lee, 1995a). This archive concentrates on damaging events over the last 200 years and was based on a systematic search through national newspapers, from 1791 to 1993. Over 4600 individual *records* (locations where erosion and flooding incidents have resulted in significant impacts) were identified. These records correspond to over 1300 separate *events* (defined as a collection of related impact records generated by the same sequence of initiating events and occurring over a similar time period). By assessing the cumulative effects of the reported incidents it was possible to classify each event according to the overall magnitude of the impact. Because of the enormous variety of impacts this was a subjective procedure, although a range of indicative criteria were used to guide the classification process (table 9.1). The average frequencies of flood events per decade over the last two centuries (1026 separate flood events were recorded) is as follows:

Table 9.1 Indicative criteria for the classification of significant events (after Lee, 1995a)

Magnitude of impact	Indicative criteria	
	Localized	*Widespread*
Minor event, Class 1	Individual towns and villages suffer flooding, with no more than ten houses inundated and four houses destroyed; flooding of minor roads. Localized erosion, including damage to bridges	National flooding of agricultural land and infrastructure; regional traffic disruption. Few communities affected
Moderate event, Class 2	Intense, localized damage in towns or villages; may involve up to five dead. Considerable local disruption, with financial hardship to a few	A region's towns and villages suffer flooding, with more than 2000 houses flooded. Damage is not intense or widespread within a community
Severe event, Class 3	Considerable localized damage in towns and villages; may involve up to 15 dead. Event may involve lengthy period of inundation and severe damage to individual properties, widespread evacuation and emergency relief	A region may experience setbacks to industry, financial hardship to a few and financial setbacks to thousands. Cities may be severely inundated in a number of districts; local towns and villages badly affected. Up to 6000 houses flooded
Major event, Class 4	Almost complete desolation and destruction to a community; may involve up to 30 dead and widespread destruction of property. Evacuation and disaster relief aid required	Considerable damage to region's cities, towns and villages; may involve over 30 dead. Widespread setbacks and financial hardship to thousands of people. More than 10 000 made temporarily homeless.

- Class 1 events (Minor) 24 per decade
- Class 2 events (Moderate) 13 per decade
- Class 3 events (Severe) 5 per decade
- Class 4 events (Major) 1 per decade

Over the same period, a total of 285 coastal erosion events were recorded, predominantly minor to moderate in estimated magnitude.

The breakdown of *event* frequency and magnitude for each decade since 1700 has been shown in figure 2.10(d), where it was compared to recorded frequency distributions of landslides and debris flows. The pattern reveals a rapid increase in the frequency of events up to the 1950s, after which the number of events has remained fairly constant. This does not imply that flooding and coastal erosion have become more frequent but, rather, that society has become more vulnerable. The factors influencing this pattern are likely to be highly complex, but are likely to include the rapid spread of development into vulnerable locations throughout the nineteenth and twentieth centuries, resulting in an increase in the chance of a damaging event. Urban growth has significantly altered surface characteristics and hydrology. The transformation of agricultural land to a housing estate, with efficient artificial drainage and impermeable concrete or tarmac surfaces, profoundly influences the quantity and rate of run-off, increasing both. Events that would have had a minor impact in the past could now affect many sectors of the economy, because of the concentration of resources and infrastructure in vulnerable locations.

At the same time there has been significant institutional and structural responses, particularly to the major flood disasters of 1947 and 1953. Improved flood warnings and defence schemes have resulted in a reduction of risk in protected areas. For example, since the East Coast floods of 1953 there has been a major programme of sea defence and flood warning improvements, which prevented a repetition of the disaster when an even higher storm surge occurred on 11–12 January 1978. It should be noted, however, that defences do not eliminate the risk of a damaging event. Flood defences in Perth, Tayside were raised in 1974, to a design level that had not been exceeded since 1814. However, in January 1993 the defences were overtopped and around 1500 properties affected in and around the city.

9.3 The Impact on Society

9.3.1 Loss of life

Since 1800 over 1000 people have died as a result of flood events. Many lost their lives in the following disasters: the Irish Sea flood of 6–7 January 1839 (115 deaths in Liverpool – Lamb, 1991); the collapse of the Dale Dyke Dam in March 1864 (244 deaths in the Sheffield area; Anon., 1864; Youdale, 1989); the Pennine floods of November 1866 (22 deaths – Anon., 1866); the floods of July 1875 in Wales and the Midlands (31 deaths, including the collapse of the Carne Dam in Wales – Anon., 1875a,b; table 9.2); the Louth floods, Lincolnshire, of May 1920 (22 deaths – Ballard, 1922); the collapse of the Eiglau Dam, Wales in November 1925 (16 deaths, Anon., 1925); the Thames floods in London, January 1928 (14 deaths – Mirlees, 1928); the Lynmouth floods in Devon in August 1952 (34 deaths – Delderfield, 1974); the East Coast floods of 31 January and 1 February 1953 (307 deaths – Institution of Civil Engineers, 1953: Steers, 1954); the Somerset and Devon floods of July 1968 (seven deaths – Anon., 1968; Hanwell and Newson, 1970); and the East Coast floods of January 1978 (26 deaths – Anon., 1978). Perhaps the greatest disaster over the last millennium was the Severnside coast floods of 1606, when about 2000 people drowned as sea defences were overtopped.

Fatalities associated with landslides and coastal erosion events are, fortunately, rare (Jones and Lee, 1994). Most accidents are the result of unexpected cliff falls on the coast. The Dorset coast has had a series of recent tragedies. In 1977 a school party were studying the geology of Lulworth Cove on the Dorset coast, when they were buried beneath a sudden rockslide; a schoolteacher and two pupils were killed. At Swanage, a schoolboy on a field course was seriously injured by a rock fall in February 1975, and a year later a young boy was killed by a falling rock. In 1971, a nine-year-old girl was hit on the head by falling rock at Kimmeridge, and later died of her injuries. In 1979 a woman, sunbathing on the beach near Durdle Door, was killed when a 3 m overhang collapsed. These incidents, and others, led the Chief Inspector of Wareham Police to coin the phrase 'killer cliffs', highlighting the serious danger that rock falls and landslides posed to tourists and educational parties (Jones and Lee, 1994). Landslides on inland slopes have killed few people. The single, tragic exception occurred at Aberfan, South Wales in October 1966, when part of a colliery tip collapsed and flowed into the village below. Twenty houses and a school were overwhelmed and 144 people died.

Table 9.2   Fatal dam failures in Great Britain (after Charles, 1992)

Date	Site	Comment
1799	Tunnel End, Marsden	Overtopped during floods; one dead
1810	Swellands, Huddersfield	Failure probably due to under seepage; Colne Valley flooded with five dead
1835	Whinhill	Overtopped during floods; 31 dead
1841	Welsh Harp	Two dead; overtopped during floods
1842	Glanderston	Overtopped during floods; eight dead
1848	Darwen, Blackburn	Dam failure during heavy flood; 12–13 lives lost
1852	Bilberry, Holmfirth	Settlement caused by internal erosion led to overtopping and collapse; 81 dead
1863	Dale Dyke, Sheffield	Dam breached during first filling of reservoir; 244 lives lost and extensive property damage in Sheffield
1870	Rishton	Three dead; unknown cause
1875	Carne, Wales	Twelve dead; failure due to internal erosion
1875	Castle Malgwyn	Two dead; overtopped during floods
1910	Clydach Vale	Overtopped during floods; five dead
1925	Skelmorlie, Largs, Scotland	Overtopped and breached during flood caused by release of water from flooded quarry; six dead
1925	Coedty, Dolgarrog, Wales	Overtopped and breached during flood caused by collapse of concrete Eiglau Dam; 16 dead

The loss of life associated with other hazards – soil erosion, sedimentation, river channel instability and so on – is unknown, but likely to be insignificant in comparison. However, channel scour and erosion around the piers of a railway bridge at Glanrhyd, Dyfed led to the bridge collapsing under the weight of a train in October 1987; four people died (Anon., 1987).

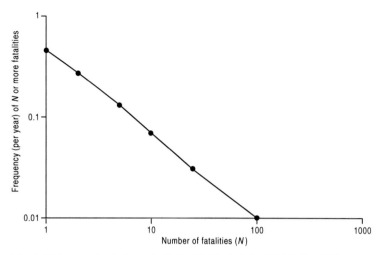

Figure 9.1 A *F/N* (frequency/number) curve for flood-related deaths in Great Britain since 1899.

The societal risk associated with these processes (i.e. the overall risk to society) can be measured in terms of the cumulative frequency of all events (F) that lead to N or more fatalities. These data are usually plotted on so-called *F/N* curves, which allow a ready comparison between different risks. Figure 9.1 presents an *F/N* curve for flooding, as derived from the historical database and, hence, covering the period of around 1899 to the present day. The curves indicate that flood events have killed ten or more people, on average, every 10–25 years, and that in most years a flood event has occurred which results in at least one fatality.

9.3.2 Economic losses

The loss of life is, however, small compared with other environmental hazards, such as poor driving conditions, extreme cold and rough seas. The true significance of these processes lies with the potential for large-scale, extremely costly impacts, spread across many sectors of the economy. The potential costs include:

1   direct damages, caused by the effects of erosion or the physical contact of floodwater with properties and their contents;
2   indirect damages, arising as a consequence of direct damage, including traffic disruption, loss of production, evacuation costs and so on; and

3   intangible damages, ranging from anxiety and stress to ill health related to the general inconvenience caused by the event.

Flooding – flash floods in upland areas, dam failures, lowland river floods and tidal floods – is the most dramatic and costly problem for society. Using a combination of nominal losses and the historical frequency of events of different magnitude, Lee and Meadowcroft (in press) have estimated the *average annual loss* from flood events, over the last 100 years, to be around £130 million. This figure disguises the enormous potential losses associated with very rare events, such as a repeat of the 1953 floods or a failure of the Thames Barrier, accompanied by widespread river flooding, when annual losses could exceed £10 billion (Clement, 1995). The average annual losses have been partly calibrated against the insurance industry's gross incurred claims for flood damage in 1998 (ABI, 1999). Domestic flood damage claims were for £118 million in 1998, with commercial claims probably in the order of £45 million. If a further £40 million is added to cover deductibles and uninsured losses (the insurance industry estimated that one in four properties damaged in the Easter floods of 1998 did not have building contents cover), then it is likely that the total losses incurred in 1998 were in the order of £200 million.

Examples of particularly distressing and costly events can be found throughout the historical record for many parts of Britain. Flash floods can occur in many catchments throughout Britain, from mountain rivers in the Scottish Highlands (table 9.3) to seasonal streams in chalk areas (e.g. the Chichester floods of 1994; Taylor, 1994). These floods are often the result of extreme rainfall events; discharges may be exceptionally high and difficult to predict. Extreme rainfall events can occur during major thunderstorms or when fronts are stationary. Occasionally, the two processes can combine and form an exceedingly rare occurrence, such as the Lynmouth floods of August 1952, which produced a discharge of 511 m^3 s^{-1}, that had only been exceeded twice on the Thames (draining an area 100 times larger) since 1883. A 'wall of water' swept down the steep channel, as temporary dams formed by trees and boulders were breached (Marshall, 1952; Kidson, 1953). In the town of Lynmouth there was widespread devastation; 25 adults and nine children were killed, 90 houses destroyed and 130 cars swept away. The total damage caused by the disaster has been estimated at £9 million (Newson, 1975).

One of the most disastrous flash floods this century was on the margins of the Lincolnshire Wolds, at Louth, on 29 May 1920. Up to 153 mm of rain fell in the small chalk catchment, eroding fields and creating 18–30 m wide torrents in normally dry valleys. The River

Table 9.3 Examples of flash flood events in the Scottish Highlands (from Nairne, 1895 and later sources)

Date	Location	Comments
3–4 August 1829	Moray district	Around 90 mm of rain; the River Findhorn rose 15 m above its normal level, causing immense damage; bridges swept away, crops and farms destroyed or ruined by deposited gravels. Numerous families left destitute and damage estimated at £20 000 (1829 prices). Severe floods on the Nairn and Spey; 'great landslips' occurred, farms swept away. In Spey valley damages estimated at over £37 000, plus countless livestock and several lives lost
27 August 1829	Inverness district	Considerable flood damage in Inverness; crops flattened, numerous bridges lost, mills and homes damaged. Estimated as several thousand pounds damage
24–26 January 1849	Inverness district	The 'Inverness Flood'; most disastrous flood in the NW Highlands. Bridges lost at Aberchalder and Forst Augustus, Caledonian Canal breached. In Inverness the stone bridge was lost and a third of the town flooded by the combined waters of the Ness and the canal. Immense damage, but no loss of life
30 January – 1 February 1868	Inverness district	Farms damaged, crops lost, bridges swept away throughout district. Inverness flooded, with extensive damage to property
29 January 1892	Strathglass, Strathspey	Great, extensive flood following unpredicted snowfalls for ten days. Damage extensive, especially in Strathglass, Bonar Bridge and Strathspey, but no fatalities. Railways washed away on Skye
25 May 1953	Lochaber, Appin and Benderloch	Road bridges destroyed, disruption to road and rail traffic; extensive damage to forestry property through floods and landslips. In Argyllshire damage to roads estimated at £130 000
30 July 1956	Cairngorm and Moray	Houses flooded and bridges damaged throughout region, especially around Forres. Main railway line from Inverness washed away. Livestock swept away. Extensive erosion and deposition of gravels on agricultural land. 72 h maximum rainfall of 250 tmm
17–18 August 1970	Moray	72 h maximum rainfall of 150 mm. Extensive damage to roads and bridges; agricultural land flooded and covered by gravels

Ludd rose 5 m in 15 minutes and a 60 m wide flood wave carrying about 140 m³ s⁻¹ swept through the small town of Louth. Twenty-two people were drowned, buildings demolished, 1250 made homeless and over £100 000 of damage caused.

The Great Till flood of 16 January 1841 also occurred in a chalk-dominated catchment, when a combination of melting snow and frozen ground resulted in a flash flood from the downlands of Salisbury Plain. The River Till burst its banks at Shrewton and rose to 2.5 m above normal level. Three people drowned and 200 were left homeless. The flood destroyed 72 houses and caused an estimated £10 000 worth of damage in the villages of Tilshead, Orcheston, Shrewton, Maddington and Winterbourne Stoke (Cross, 1967).

Seasonal flooding presents a recurrent problem in many lowland river valleys throughout Britain. Some of the worst events include the lowland floods of March 1947, which affected rivers throughout South Wales and much of England. The resulting damages were probably in excess of £500 million (at current prices). In recent years, the Perth floods of January 1993 caused an estimated £18 million of damage, mainly on the North Muirton housing estate, where 780 homes were affected as floodwaters reached a depth of 2 m (Babtie et al., 1993; table 9.4). The 1998 Easter floods affected over 4200 properties in England and Wales, with over 2000 properties flooded, two dead and 150 people treated for hypothermia in Northampton alone (Agriculture Committee, 1998). At the time of writing, the widespread flooding of November 2000 has caused considerable damage across southern England, parts of the Midlands, Yorkshire and eastern Scotland.

It has been estimated that over 5% of the population lives in areas below 5 m AOD and, hence, is at risk from coastal flooding. Low-lying coastal land is extremely vulnerable to events which involve either overtopping or breaching of sea defences, especially because of the speed of flooding in such circumstances. Tidal floods can cause extensive damage and distress in the affected area. During the 1990 Towyn flood, seawater overtopped and breached the flood defences along the Clywd coast, inundating 2800 homes in towns and villages from Pensarn to Rhyl, with estimated damages of around £70 million. The floodwater was deepest at Towyn, where 1000 people were evacuated to neighbouring towns. Although there were no drownings, it has been suggested that as many as 50 people may subsequently have died from flood-related trauma (Welsh Consumer Council, 1992).

The low-lying areas of the East Coast of England are particularly vulnerable to tidal flooding. At least seven floods have had major impacts since the thirteenth century (Jensen, 1953), including the 6 January 1928 floods on the Thames, when 14 drowned in London

Table 9.4 Direct damage for flooding in and around Perth, February 1990 (after Falconer and Anderson, 1992)

	Overall direct damage cost (£)	Percentage of total
Agricultural damage		
Tay catchment	640 000	20
Floodbank reinstatement	375 000	12
Other damage	50 000	2
Earn catchment		
Subtotal	£1 065 000	
Building samage		
Perth	735 000	23
Outside Perth	425 000	14
Subtotal	£1 160 000	
Public and other authorities		
Scotrail	193,000	6
Scottish Hydro Electric Plc	250 000	8
Perth and Kinross District Council	80 000	2
Tayside Regional Council		
Water services	5 000	–
Roads	381 000	12
Tay River Purification Board	1 000	–
SSPCA	4 000	–
Tay District Salmon Fisheries	20 000	1
Subtotal	£3 159 000	29
TOTAL	£3 159 000	

basements and 4000 were made temporarily homeless. Samuel Pepys described the great flood of 1663: 'there was last night the greatest tide that was ever remembered in England to have been in this river [the Thames], all Whitehall having been drowned'. The Anglo-Saxon Chronicles record a great flood in November 1099, when 'the great sea-flood came up and did so much harm that no man remembered its like before'. However, the greatest sea flood, in terms of the resulting damage, occurred on 31 January 1953 and inundated over 800 km² of eastern England. Over 300 died and extensive damage was caused by what was then the highest tide level ever recorded. Thousands of homes and factories were damaged at an estimated cost of £5 billion (at current prices; Willis Faber and Dumas, 1996); there were 1200 breaches in the sea defences (Steers, 1953; ICE, 1954). The situation was even worse in the Netherlands, where 1600 km² was flooded and 1800 lives lost (Volker, 1953).

By contrast with the spatially extensive problems associated with flooding, *coastal erosion* tends to create more site specific or localized difficulties. Even so, it can still pose a threat to development, as illustrated by the major coastal landslide at Holbeck Hall, Scarborough in June 1993, which is likely to have resulted in excess of £3 million of damage and repair works. Coastal recession leads to land loss. Although individual failures generally tend to cause only small amounts of cliff retreat, the cumulative effects can be dramatic. For example, the Holderness coast has retreated by around 2 km over the last 1000 years, with the loss of at least 26 villages listed in the Domesday survey of 1086 (Valentin, 1954; Pethick, 1996). On parts of the north Norfolk coast there has been over 175 m of recession since 1885 (Clayton, 1980, 1989); county archives show that 21 coastal towns and villages have been lost since the eleventh century. Rapid recession has also caused severe problems on the Suffolk coast, most famously at Dunwich, where all that remains is a fragment of the cemetery.

The public perception is that parts of Britain are being rapidly lost to the sea, and that properties still regularly fall over cliff tops. The reality is somewhat different, primarily because of the effectiveness of coast protection measures built over the last 100 years or so. The average annual loss of land due to cliff recession and coastal landsliding is probably less than 25 ha (High Point Rendel, in press). Examples of property being lost are actually quite rare, probably no more than 1 per year, on average and often the buildings have been declared unsafe by the local authority and demolished (at the owner's expense!) before they fall over the cliff.

Along parts of the South Coast, landsliding has been widely reported as affecting many coastal communities from Kent to Cornwall, including Sandgate, Peacehaven, Ventnor, Barton-on-Sea, Christchurch, Swanage, Portland, West Bay, Charmouth, Lyme Regis, Torbay and Downderry (Lee, 1993a). The most extensive coastal landslide problem in Great Britain is at Ventnor, on the Isle of Wight, where the whole town has been built on an ancient landslide complex (Lee and Moore, 1991). Although present-day coastal retreat is minimal, long-term erosion has helped shape a belt of unstable land that extends almost 1 km inland. Contemporary movements within the town have been slight. However, because movement occurs in an urban area with a permanent population of over 6000, the cumulative damage to roads, buildings and services has been substantial. Over the last 100 years about 50 houses and hotels have had to be demolished because of ground movement.

Inland landslide activity mainly involves the reactivation of pre-existing failures and rarely involves dramatic fast-moving events. Even when large displacements occur the rate of movement tends to be relatively slow. Landslides do, however, pose a very real threat to the construction and development industry (Jones and Lee, 1994). The cumulative effects of slow movement are known to have caused considerable damage to buildings, structures and infrastructure. Damage due to slope instability can lead to expensive remedial measures or, where repair is considered uneconomic, the abandonment and loss of property. The permanent closure of the A625 Manchester–Sheffield road at Mam Tor in 1979, and the decision not to reopen the railway link to Killin following the Glen Ogle rockslide in 1965, are amongst the best known examples.

Deposition within river channels can lead to serious maintenance and operational problems. Annual maintenance dredging costs in excess of £1 million are incurred at Harwich and Liverpool. At Kings Lynn, the approaches have to be resurveyed every two weeks, with navigation buoys repositioned up to 100 times a year. In 1993, for example, British Waterways spent over £3 million on dredging, involving the removal of 300 000 tonnes of material from the canal network. Sedimentation in reservoirs can lead to significant losses in water storage capacity. It was shown recently that loss of capacity in a sample of southern Pennine reservoirs varied between 4% and 75% in around 100 years since construction (Butcher et al., 1992). Elsewhere, records provided by Northumbria Water indicate a variable pattern, with loss of capacity generally less than 15% over 100 years. Deposition within the channel can, of course, reduce its capacity and lead to flood problems, as reported for the River Spey in Grampian and the Findhorn in Highland. Sear and Newson (1992) estimated that an annual total of £7 million is spent on sediment-related maintenance problems along rivers in England and Wales. In recent years concerns have been expressed over the impact of hillslope erosion on the discoloration of water supplies, increasing the cost of water quality treatment.

Other processes such as hillslope erosion, wind erosion and channel migration can lead to notable problems for affected landowners, and can lead to difficulties where infrastructure and services cross vulnerable areas. The implications of these problems are easy to dismiss as trivial; the following examples should serve to demonstrate that they can lead to serious problems. Erosion of upland peat is seen as a serious threat to many valued landscapes, especially in the Pennines (Labadz et al., 1991). Soil erosion and mudflood problems in the South Downs during October 1987 probably resulted in £0.75 million of damage, especially in and around Rottingdean (Robinson and Blackman, 1990).

Wind erosion, both inland and in coastal sand dunes, has resulted in dramatic events. Perhaps the most remarkable inland event occurred during exceptionally severe storms between 1570 and around 1668, in Breckland, East Anglia. The town of Santon Downham was gradually engulfed by moving sands by around 1630; farmhouses were buried and later exposed as the sands moved on. Lamb (1991) suggests that between 50 and 100 Mm³ of sand was involved in the 'sand floud' or 'wandering sands', possibly derived from exposed sandy soils around Lakenheath. Soil erosion by wind can damage crops and reduce productivity by removing seeds, exposing roots and blasting leaves, and by reducing soil quality. The dust storms can reduce visibility and block roads, ditches and fences. Although many farms in Britain may be affected by a degree of wind erosion, with its cumulative effect on soil quality, notable off-farm problems tend to arise only in unusual circumstances. In March 1968, for example, a series of soil blows across parts of Lincolnshire led to a range of local impacts (Robinson, 1969). Many roads were partially blocked by wind-blown material and traffic was disrupted. Clearing operations in Lindsey alone cost £4000. Ditches and drains were filled with sediment: it cost one drainage board £5000 to clear ten drains. Many farmers had to clear ditches on their own land, at their own expense: the average cost was estimated to be approximately £5 per 20 m, and in the Isle of Axholme alone the cost may have been £17 500. Also on farmland, productivity was reduced in places by uncovering or removal of seeds (e.g. barley, peas and beet), and by the 'scorching' of leaves and root-exposure of winter wheat plants.

Migration of sand dunes is not a significant problem in Great Britain, but it once was. There is considerable historical evidence to suggest that wind-blown sand was a major hazard, especially during the fifteenth to seventeenth centuries (table 9.5). Perhaps the most memorable series of events was the Culbin Sands disaster of 1694 and following years, which has been described by Jones (this volume, section 3.4). The loss of land on the fertile Moray plain would probably have been valued at £15 million at current prices, with buried property and lost crops probably raising the overall total to around £25 million (Lamb, 1991). In addition, the loss of natural coast protection that had been provided by the dunes before they migrated inland led to the destruction of the nearby town of Findhorn during a storm around 1702.

Table 9.5 Examples of major sand migration events in Great Britain (after Lee, 1995a)

Date	Site	Comment
1316	Kenfig, near Port Talbot	Storms causing sand dune movement closed the medieval port of Kenfig. Further events between 1344 and 1480 finally buried the former Roman coast road
1385	Harlech	Around this date, sand dunes formed, enclosing and protecting the flat area known as Morfa Harlech, closing the medieval port of Harlech
1401–13	Forvie, Grampian	Medieval town buried by sands (now 30 m high dunes). The Forvie dune advanced 50–250 m to the north during the 1413 storm and advanced a further 200 m before the end of the fifteenth century. The storm corresponded with extreme low tides
1600	Rattray, Grampian	From 1600 to 1720 the inlet of Strathbeg was buried by sand; it is now the Loch of Strathbeg. Rattray harbour was choked with sand
1663	Nairn, Grampian	Moving sands threatened to cover the town
1676	Culbin Sands, Grampian	Culbin estate covered by blown sands to a depth of 0.7 m. The source was coastal dunes, where marram grass had been eaten by sheep and cattle
1694	Culbin Sands, Grampian	In the autumn of 1694, 16 fertile farms covering 20–30 km^2 were overwhelmed in a violent storm. The whole area was buried by up to 30 m of sand
1697	North Uist	Archaeological site buried by drifting sand, carried from shore
22 October 1702	Findhorn, Moray Firth	A severe drift of loose sand in the Culbin area blocked the River Findhorn and forced it to change its course
1739	Sefton	The village of Ravenmeols was buried by a great sandstorm
1794	Happisburgh, Norfolk	The church was buried by sand in a storm
26 December 1862	Happisburgh, Norfolk	The remains of the ancient village of Eccles, which had been buried in sand in the seventeenth and eighteenth centuries, were exposed by the exceptionally strong winds
1870s	Alnmouth	Between 1866 and 1897, a belt of sand dunes developed on the south side of the estuary

9.4   Management Responses: Individuals and Private Enterprise

A basic and long-standing principle of British law is that individuals have the right to protect their own property, under common law. Hence, the primary responsibility rests with the landowner, not with the state. However, as will be described below, common law rights have been altered and reduced over time by statute law to allow state intervention in the interests of the common good. Individuals do not have to exercise their rights, although case law has indicated that landowners or occupiers have a general duty to their neighbours to take reasonable steps to remove or reduce hazards if they know of the hazard and of the consequences of not reducing or removing it.

For much of the last 1000 years, individuals or private businesses have either avoided high-risk areas, accepted the losses as the price to pay for living and working in such areas, or sought to 'improve' the conditions through engineering works. Maintenance, repair and clean-up are often a central element of most strategies for dealing with natural hazards (Lee, 1995b). Insurance has become available for mitigating the losses associated with flooding or landslip (but excluding landslide losses caused by marine or river erosion). Occasionally, compensation has been sought through litigation.

The history of flood defence and protection against coastal erosion is inseparable from that of wetland reclamation and land drainage. Throughout the last 1000 years, risk reduction has tended to go hand-in-hand with increased agricultural productivity and profitability, and has often been associated with private enterprise. Where public expenditure was involved, it was generally based on co-operation at a local level, rather than direct state funding. These patterns have changed over the last century, to the extent that state funding of urban flood alleviation or protection against coastal erosion has become the dominant response, at the expense of land drainage works (Penning-Rowsell et al., 1986).

An extensive network of drainage channels, dykes and sea walls, many of which had been built by the Romans, predated the start of this millennium. For example, Roman embankments along the River Medway survived until the eighteenth century (Harrison and Grant, 1976). Purseglove (1989) noted that a drainage engineer from the Somerset Levels, Girard Fossarius (Gerard of the Drain) was listed in the Domesday Book of 1086. By this time, the Church had become an important force in land drainage and continued to reclaim marshland for agricultural use until the dissolution of the monasteries in 1530. For example, the monks of Furness reclaimed the Walney marshes,

Cockersands Abbey reclaimed parts of the Fylde and the Bishop of Durham instigated extensive drainage and flood defence works along the northern shores of the Humber estuary. Between 1150 and 1300 there was considerable reclamation around the Wash by local communities. Hoskins (1955) notes that the newly claimed land was divided into lots, each of which carried obligations to keep a certain length of ditches and dykes in repair. He also cites the example of one Thomas Flower who, in 1439, failed to repair his section of Wisbeach fendike and, as a result, allowed some 12 000–13 000 acres to flood.

By the thirteenth century, the men of Dunwich on the Suffolk coast had begun their attempts to prevent coastal erosion. In 1222 the burgess of Dunwich sent a petition to the king for aid in 'enclosing' their town (i.e. building a sea wall). No help came and the locals were left to build the defences (probably a sand and shingle bank, faced with clay and fronted by bundles of sticks; Parker, 1978) at their own cost. At Lyme Regis, The Cobb was constructed as a breakwater around 1250. In 1586 Sir Francis Walsingham reported that 'an exceeding number of great piles' had been constructed to protect the town; these piles were probably a series of oak-groynes arranged both at right angles and parallel to the shore, filled with shingle banks (Fowles, 1982).

In the late sixteenth century, the pattern of wetland reclamation and, hence, flood defence was transformed by the arrival of new technologies from Holland (Purseglove, 1989). Extensive reclamation works involving ditches and dykes (flood embankments) were carried out to provide high-quality agricultural land, often at the expense of common pasture lands or marshes. The operations generally involved a drainage engineer (an 'undertaker') and a private investor (an 'adventurer'); a large proportion of the reclaimed land rewarded both. In the 1620s Cornelius Vermuyden reclaimed Hatfield Chase, south of the Humber, for Charles I, in return for a third of the land drained. In 1630 the Bedford Level Corporation employed Vermuyden to undertake the drainage of the Fens. The Old Bedford River was constructed in 1637. The New Bedford, or Hundred Foot, River was completed after the civil war, in 1651. The two parallel channels and associated earth bunds enclose a floodplain, the Ouse Washes, which carries the floodwaters in winter, thereby protecting the adjacent agricultural land.

Peat shrinkage, however, led to a renewal of the flooding problems across the Fens within 25 years of Vermuyden's works. New technologies, pumps driven by windmills and later steam, helped secure the land drainage improvements. Further drainage achievements followed the invention of underdrainage with clay tile pipes in the late eighteenth century by Joseph Elkington. Elsewhere, Capability Brown and Repton 'improved' many estates by draining marshes and valley bottoms, and creating landscaped lakes. In 1811 William Maddocks reclaimed the coastal marshes of the Treath Mawr, North Wales, using drainage ditches and an embankment which carried the road and railway to Porthmadoc. Further reclamations continued through the nineteenth century, as a rapidly increasing urban population needed to be supported by more efficient agricultural production. Almost 2000 ha of saltmarsh in the Ribble estuary were reclaimed in the nineteenth century. Some 32 000 ha around the Wash have been reclaimed since the seventeenth century; and up to 85% of the intertidal area of the Suffolk estuaries has been reclaimed since the twelfth century (Beardall et al., 1988).

During the latter half of the nineteenth and early twentieth centuries, many local authorities constructed sea walls (often as a way of relieving unemployment) which combined the functions of coast protection and a promenade for the expanding tourist industry. In Scarborough, North Yorkshire, sea walls were constructed in the 1880s and 1890s to provide protection against flooding and, along with landscaping and drainage works, to stabilize the eroding coastal cliffs. The Undercliff Drive promenades and sea wall in Bournemouth were opened in 1911, with 29 precast concrete groynes installed between 1937 and 1939 to maintain beach levels (Lelliott, 1989). The expansion of the railways was accompanied by the construction of flood embankments and sea walls to protect the lines, as along the North Wales coast.

In such a short summary, it is difficult to convey the very scale and extent of the land drainage, wetland reclamation and coastal defence works that have been undertaken by landowners and private enterprise since the sixteenth century. A total of 91 250 ha of land claim has occurred within British estuaries, mainly over the course of this millennium (Davidson et al., 1991). Some form of flood defence works to secure the new land has accompanied all these operations. Reclamation and drainage works continued throughout the twentieth century, albeit at a lower intensity, reflecting a gradual decline in the agricultural sector and the increasing awareness of the conservation importance of natural wetlands.

9.5 Management Responses: State Intervention

Although, as stressed earlier, the ultimate responsibility for managing risks rests with individual property owners, the state has gradually acquired a key role in addressing a number of specific problems (see, e.g., Lee 1993b). These include:

- the provision of publicly funded flood defence works and coast protection works to prevent erosion or encroachment by the sea
- the provision of flood warning systems
- funding and coordinating the response to major events
- controlling development in areas at risk and minimizing the impact of new development on risks experienced elsewhere, through the land use planning system

There are a number of reasons as to why state intervention has become increasingly significant over the last millennium. As Penning-Rowsell et al. (1986) note, the scope for individuals to reduce their own exposure is generally limited to pragmatic measures such as 'floodproofing' or insurance. The cost and complexity of flood and coastal defence works is generally beyond most property owners, with the exception of major companies (e.g. Railtrack) or landowners. Indeed, it is often neither feasible nor desirable to attempt to protect a single property. To do so would inevitably lead to a patchwork of defence structures of different condition, standards and performance. For example, a specific national agency, the Environment Agency (EA), has been empowered to exercise supervision over all matters related to flood defence in England and Wales. State involvement also has a social welfare element. For example, reduced flood damage should help promote greater prosperity by ensuring the security of property, a healthy workforce and efficient business. There is also a need to balance the pressures for reducing the risks faced by communities and obligations to take into account the interests of other groups, such as conservation bodies and fisheries interests. The evolution of Statute law has, therefore, introduced (Lee, 1995b):

- consenting arrangements that ensure that management measures do not affect other interests or increase the level of risk elsewhere
- provisions to ensure the conservation and enhancement of landscape and nature conservation features, involving the protection of designated sites and areas of national and international importance
- consultation arrangements between key interest groups whose interests may be affected by risk-management measures

9.5.1 Flood defence and coast protection

The origins of institutionalized flood defence date back to the establishment of the Commissioners of Sewers in 1427 (the first commission had been set up in Lincolnshire by Henry de Bathe in 1258) and the Bill of Sewers in 1531 (a sewer was a straight cut, or drain). The commissions had responsibility for land drainage and reclamation from the sea, and were answerable to central government. The system survived until the 1930s. The modern administrative framework in England and Wales developed from the Land Drainage Act 1930, which included defence against sea water within the definition of drainage. Land drainage and flood defence was funded by precept on county councils. Subsequent changes include the privatization of the water industry and the establishment of the National Rivers Authority (NRA) by the Water Act 1989, whose flood defence powers were defined in the Water Resources Act 1991. The NRA has since been reorganized and flood defence is the responsibility of the Environment Agency. Local authorities and internal drainage boards also have long-standing land drainage and flood defence powers, as currently defined by the Land Drainage Act 1991. In Scotland, local authorities have limited flood defence powers, under the Flood Prevention (Scotland) Act 1961.

By contrast, there were no general statutory powers to protect the coast against erosion before 1949. However, many authorities, such as in Scarborough, had provided defences under general local authority powers or local Acts. Following widespread deterioration of the defence during the Second World War, it was recognized that private owners lacked the resources to carry out the necessary repairs. The Coast Protection Act 1949 thus gave local authorities powers to carry out works, under general supervision by central government (through the Ministry of Agriculture, Fisheries and Food; MAFF), to prevent erosion or encroachment by the sea.

The various Acts enable the relevant operating authorities to undertake defence measures and enable the government to offer financial support for specific works ('grant-aid'). A key feature of the powers is that they are permissive rather than mandatory; the operating authorities are not obliged to carry out works. This clearly limits the role of the state to only providing defences that are deemed to be in the national interest. However, the subtle distinction can cause considerable public misunderstanding and frustration. It also recognizes that complete protection is impossible: 'a balance has to be struck between costs and benefits to the nation as a whole. For example, to attempt to protect every inch of coastline from change would not only be uneconomic but would work against the dynamic processes which determine the coastline and could have adverse effect on defences elsewhere and on the natural environment' (MAFF, 1993a).

Protection of life is the primary focus of government flood and

Table 9.6   Indicative standards of flood protection (after MAFF, 1993b)

Current land use	Indicative standards of protection (return period, years)	
	Tidal	*Non-tidal*
High-density urban, containing a significant amount of both residential and non-residential property.	200	100
Medium-density urban, lower density than above – may also include some agricultural land	150	75
Low-density or rural communities, with limited number of properties at risk. Highly productive agricultural land	50	25
Generally arable farming with isolated properties; medium-productivity agricultural land	20	10
Predominantly extensive grass, with very few properties at risk, low-productivity agricultural land	5	1

coastal defence policy (MAFF, 1993a). The order of priority for grant-aid is as follows:

- flood warning systems
- urban coastal defence
- urban flood defence
- rural coastal defence and existing rural flood defence
- new rural flood defence schemes

These priorities are not prescriptive and grant-aid decisions are subject to rigorous appraisal procedures (MAFF, 1993b). In England and Wales, schemes (especially those that are grant-aided by MAFF or the National Assembly for Wales) must be technically sound, environmentally acceptable, economically viable and cost-effective. The standard of protection provided by defences varies with the nature of the land use in the area at risk. MAFF, for example, expresses indicative standards of protection in terms of the flood return period for five subjectively expressed current land use bands (table 9.6). These standards are intended as guidance, not to set minima.

The existing arrangements for flood and coastal defences have been very effective in protecting vulnerable communities. For example, over the last 100 years or so, some 860 km of coast protection works have been constructed in England, to prevent coastal erosion (MAFF, 1994). However, the arrangements have tended to lead to a compart-mentalized response to erosion and flooding processes, whereas in reality their interaction is an important factor in initiating potentially damaging events. In the river environment, for example, channelization works to prevent bank erosion can lead to channel adjustments both upstream and downstream, creating increased flood risk or erosion problems for other landowners. The construction of coast protection works may lead to an increase in flood risk elsewhere, as saltmarshes, mudflats, beaches and shingle ridges, which provide natural defences to low-lying areas behind, are starved of sediment. However, in the last decade there have been significant changes in attitudes towards the use of engineering schemes to managing flooding and coastal erosion problems. The main changes have been the recognition of a need for strategic planning of flood and coastal defence issues through Shoreline Management Plans (MAFF, 1995) and the advocacy of soft engineer-ing approaches on the coastline (Pethick and Burd, 1993). Expenditure on flood and costal defence in England and Wales is indicated in table 9.7.

9.5.2 Flood warning systems

After the 1953 East Coast floods, the Waverley Committee (1954) recommended that flood warning systems should be set up so that early action could be taken in advance of future events. The Storm Tide Warning Service (STWS) was established to provide warnings of high surge tides on the East Coast. The distribution of surges is heavily biased towards the winter months when Equinox Spring Tides occur, with little surge activity experienced in the summer months. The STWS, therefore operates on a seasonal basis with 24-hour-a-day manning from 1 September to the end of April, with a general eye kept on the situation over the summer. The Meteorological Office operates the STWS on behalf of MAFF, who are responsible for ensuring that adequate warning procedures exist. On the South and West Coasts, the STWS have developed a mathematical surge forecasting model and issue messages to the EA regions, drawing their attention to any potential flood conditions.

Flood warning systems are well established for river floods in Eng-land and Wales, relying on a combination of storm weather forecasts by the Meteorological Office and river flow gauges. Powers to set up

Table 9.7 An indication of the expenditure on flood and coastal defences, in England and Wales (after MAFF, 1998)

Source	Purpose	Estimated annual cost, £M (year of figures)
Revenue Support Grant	For payment of levy to the Environment Agency (EA) and to Internal Drainage Boards (IDBs); also to local authorities for their own flood and coastal defence programmes	225.4 (1997/8)
Supplementary Credit Approvals	MAFF dispenses SCAs to local authorities to cover any remaining expenses incurred in construction of defences which cannot be met from its own resources	15 (1998/9)
Grant and Grant-in-aid	MAFF provides flood defence grant to the EA and IDBs and grant-in-aid to local authorities for construction of flood and coastal defence works	54 (1998/9)
Drainage rates and charges	Payable by farmers to fund IDB expenditure	11 (1996/7)
Private-sector contributions	Estimated by MAFF to be 1–2% of the capital programme	3
Other costs	MAFF research and development, funding the Storm Tide Warning Service, running costs and so on	6.2 (1998/99)
Estimated total		314.6

flood warning systems are set out in the Water Resources Act 1991. The EA provides a flood warning service to 960 locations in England and Wales. Flood Warning Dissemination Plans set out the role of the Agency, other organizations and the public in each area. All individuals who live or work in the high-risk area have been sent an information leaflet that summarizes the plan and warning service, and explains the yellow, amber and red coding system for the severity of the warning. Regional flood forecasting centres receive data from a local radar station and from the national network of weather radar sites. This data is used to estimate rainfall totals falling in a catchment and, together with rain gauge readings, forms the basis for providing advance warnings of flooding events to the police and local authorities. However, the service came under severe criticism after the 1998 Easter floods (see, e.g., Agriculture Committee, 1998).

9.5.3   Emergency action

Local authorities have a long tradition of undertaking works to benefit their residents, either to alleviate the suffering after a disaster or preventing problems. Today, a local authority has the permissive powers, under the Local Government Act 1972, to:

- incur expenditure which, in their opinion, is in the interests of their area or its inhabitants
- incur such expenditure as they consider necessary in an emergency or disaster involving destruction of or danger to life or property, or where there are reasonable grounds for preventing such an event
- make grants or loans to other people or bodies in an emergency or disaster

The emergency services can be involved throughout a flood event, from the preparation of emergency plans to the supervision of recovery operations (Parker, 1988). Emergency relief ensures that some of the immediate losses are spread throughout the community. The costs of the operation can be considerable; the response to the 1982 York and Selby floods cost a reported £366 000 (Parker, 1988). The EA also operates a flood emergency response, employing a workforce of 1607 people to check the integrity of defences and to operate sluices and barriers. The workforce is sized to allow the Agency to deal with the first 12 hours of a moderate-sized event. This provides the time to identify the need and mobilize external resources. For example, during the 1994 Chichester floods, personnel from five EA regions were deployed in the emergency response.

Rehabilitation costs can be eased by disaster funds or grant payments made to the local authority under a 'Bellwin' scheme, established under the Local Government and Housing Act 1989. Under this scheme the Department of the Environment/National Assembly for Wales makes available financial assistance for 85% of local authority expenditure incurred above a threshold level (DoE, 1993):

- in providing relief or carrying out immediate works to safeguard life or property, or prevent suffering or severe inconvenience
- as a result of the incident specified in the scheme
- on works completed before a specified deadline (usually within two months of the incident)

9.5.4 Control of development

The planning system (originally established in 1947), as currently defined by the Town and Country Planning Act 1990 and the Town and Country Planning (Scotland) Act 1972, aims to regulate the development and use of land in the public interest. Planning powers are exercised by local planning authorities, whose functions include the preparation of development plans and the control of development, through the determination of planning applications.

Development plans can be used to set out broad strategic policies (Structure Plans or UDP Part I's) or detailed policies (Local Plans or UDP Part II's) that establish a framework for restricting development. The allocation of land for specific types of development can be made with the need to avoid certain vulnerable areas in mind. The development control process can ensure that planning permission is refused in vulnerable areas. Typical approaches that have been used include identification of floodplain or coastal lowland areas at risk from Environment Agency flood risk maps and defining a 'set-back line' within which development could be affected by coastal erosion over a particular time period.

Prior to the mid-1980s, local planning authorities frequently viewed natural hazards such as erosion and flooding as technical problems that the landowner and developer needed to overcome, or the responsibility of coast protection authorities and drainage authorities; they were not seen to be land use planning issues (Lee, 1995c). Since the mid-1980s, there has been a notable change in perception about the way in which problems are managed. These changes reflect a growing appreciation that the past approach was not in the public interest:

- development in vulnerable locations can lead to demands for expensive publicly funded defence works
- develpment can have possible adverse effects on the level of erosion or flood risk elsewhere
- defence works can have significant adverse effects on the interests of other users of rivers or the coastal zone
- defence works can encourage further development in vulnerable areas, increasing the potential for greater losses when extreme events occur

The change in attitude also reflects concern about the possible effects of global warming and sea-level rise and, at a local level, the effects of recent major hazard events such as the North Wales floods of February 1990, the Tayside floods of 1990 and 1993, and the Holbeck Hall coastal landslide of June 1993. The government has advised planning authorities in England and Wales that it is the purpose of the planning system to 'regulate the development and use of land in the public interest' and that planners need to take into account 'whether the proposal would unacceptably affect amenities and the existing use of land and buildings which ought to be protected in the public interest' (PPG1 – DoE, 1992a; PPG12 – DoE, 1992b). Clearly, development in vulnerable areas is not in the public interest if appropriate preventative or precautionary measures have not been taken, or if they lead to significant adverse effects on the environment or other interests. The potential impacts of development proposals on the public interest are clearly land use planning issues, especially when public funds are sought later to protect against natural hazards.

In recent years the government has emphasized the need for planning authorities to take natural hazards into account, through specific policy guidance:

- PPG14, *Development on Unstable Land* (DoE, 1990), together with *Annex 1: Landslides and Planning* (DoE, 1996)
- PPG20, *Coastal Planning* (DoE, 1992c)
- Circular 30/92, *Development and Flood Risk* (DoE, 1992d)

Where development is permitted in vulnerable areas and adequate defences cannot be provided, the levels of risk to property owners can be reduced by incorporating specific floodproofing or ground movement tolerating measures into the building design. In flood-prone areas, the most effective building modifications include the following (Lee, 1995b):

1 *Minimum floor heights.* Property can be elevated above a prescribed design flood level either by structural means (stilts) or by raising the property on an earth bund. In the Royal Borough of Windsor and Maidenhead, for example, new residential properties on the Thames and Colne floodplains, built on land flooded in 1974, must have an internal ground floor level 0.15 m above the 1947 flood level (the highest flood in the twentieth century). Similar policies can apply to areas at risk from coastal flooding. Southampton City Council, for example, has specified that new houses built on low-lying land require floor slabs at 3.4 m AOD, with all car parks and highways at 3.1 m AOD.

2 *Means of escape.* Residents of single-storey properties are particularly vulnerable to flood events, as they cannot escape to safety upstairs as the floodwaters rise. The risks to such individuals can be reduced by requiring a means of escape such as a 'dormer window' to be incorporated into the building design. Swale Borough Council, for example, have specified that any new houses on land less than 5.3 m AOD should contain a means of escape at first-floor level, unless the site is protected by secondary defences. In the Romney Marsh area, Shepway District Council also require single-storey houses, in areas liable to shallow-water flooding, to have a means of escape, with no single-storey housing allowed in deep water flood areas.

New development can significantly increase the quantity and rate of run-off that reaches watercourses, through the creation of extensive areas of impermeable materials. These effects can lead to flooding when the additional water causes the channel capacity to be exceeded during rainfall or snowmelt events. Problems are often associated with culverts, bridges and other channel constrictions. The planning system is the principal mechanism for ensuring that development does not increase the risk of flooding elsewhere due to the generation of additional run-off. In England and Wales, the relationship between local planning authorities and the Environment Agency, as detailed in DoE Circular 30/92 (DoE, 1992d), is intended to ensure that potential run-off problems are considered throughout the planning process, from development plan preparation to the determination of planning applications.

9.6   Discussion

It is clear that, despite the postwar structural responses, geomorphological processes have increasingly imposed themselves upon many communities, creating frequently unexpected problems to homeowners and
businesses that were largely unaware of the risks that could be antici-
pated. This is readily apparent on the broad floodplains of Britain's
major rivers, on the soft rock cliffs of eastern and southern England,
and in the coastal lowlands of North Wales and England. It has been
estimated from the historical record that the average level of damage or
maintenance and defence needs associated with these processes probably exceeds £300 million per year (Lee, 1995a). These costs are, of
course, spread through many levels of the economy, from individuals
to industry, local authorities to national government.

In Britain, a range of responses has been adopted throughout the
last millennium, by individuals and the state, for the management of
natural hazards, including:

- *acceptance* of the risk
- *reducing* the occurrence of potentially damaging events
- *avoiding* vulnerable areas
- *protecting* against potentially damaging events

In most cases, however, the response has been complex, involving a
variety of measures adopted by different organizations at different
locations. Over time, the responses have become more sophisticated.
The prevailing solutions have reflected the nature of the problem, the
prevailing attitudes to natural hazards and level of acceptable risk, the
technologies available at the time, the availability of resources and the
statutory powers available to the individuals, interested bodies or
authorities to tackle the problems. Passive acceptance of natural hazards as 'acts of God' was replaced by 'nature to be commanded' and,
by the end of the 1990s, 'living with natural hazards' and 'working with
natural processes'. British law has evolved gradually; the present administrative framework for the management of natural hazards must,
therefore, be seen to be the product of the way in which they have
presented problems to society in the past. The framework has developed out of a long-standing need to tackle conflicts between an
individual's or a community's need for protection, the restriction of
common law rights for the general good, and obligations to take into
account the interests of other groups, such as conservation, fisheries
and recreation.

REFERENCES

Agriculture Committee 1998: *Sixth Report: Flood and Coastal Defence*. London: The Stationery Office.

Anon. 1864: Terrible calamity at Sheffield. *The Times*, 14 March, 9.

Anon. 1866: The floods in Lancashire and Yorkshire. *The Times*, 17 November, 12.

Anon. 1875a: The British floods of July 1875. *Meteorological Magazine*, 10, 97–111.

Anon. 1875b: Floods. *The Times*, 21 July, 7.

Anon. 1925: Conway Valley dam disaster. *Western Mail*, 4 November, 3, 7.

Anon. 1968: Seven killed in flood havoc. *Daily Telegraph*, 12 July.

Anon. 1978: At least 26 die in gales and flood devastation. *The Daily Telegraph*, 13 July.

Anon. 1987: Four killed as train plunges into river. *Western Mail*, 20 January, 1, 5.

Association of British Insurers (ABI) 1999: *Statistical Bulletin*. London: ABI.

Babtie, Shaw and Morton 1993: *Flooding in the Tay Catchment – January 1993*. Report to Tayside Regional Council.

Ballard, G. A. 1922: Grantham: its roads, unemployment and flood. *The Surveyor*, 62, 223.

Beardall, C. H., Dryden, R. C. and Holzer, T. J. 1988: *The Suffolk Estuaries*. Suffolk Wildlife Trust.

Brunsden, D., Ibsen, M-L, Lee, E. M. and Moore, R. 1995: The validity of temporal archive records for geomorphological purposes. *Quaestiones Geographicae*, Special Issue 4, 79–92.

Butcher, D. P., Claydon, J., Labadz, J. C., Pattinson, V. A., Potter, A. W. R. and White, P. 1992: Reservoir sedimentation and colour problems in southern Pennine reservoirs. *Journal of the Institution of Water and Environmental Management*, 6, 418–31.

Charles, J. A. 1992: Embankment dams and their foundations: safety evaluation for static loading. Keynote Paper, *International Workshop on Dam Safety Evaluation*, Grindelwald, Switzerland.

Clayton, K. M. 1980: Coastal protection along the East Anglian coast. *Zeitschrift für Geomorphologie*, Supp., 34, 165–72.

Clayton, K. M. 1989: Sediment input from the Norfolk cliffs, eastern England – a century of coast protection and its effect. *Journal of Coastal Research*, 5, 433–42.

Clement, D. 1995: Property insurance and flood risk. *Proceedings of the 30th MAFF Conference of River and Coastal Engineers*. London: MAFF, 2.21–7.

Cross, D. A. E. 1967: The Great Till Floods of 1841. *Weather*, 22, 430–3.

Davidson, N. C., d'A Laffoley, D., Doody, J. P., Way, L. S., Gordon, J., Key, R., Drake, C. M., Pienkowski, M. W., Mitchell, R. and Duff, K. L. 1991: *Nature Conservation and Estuaries in Great Britain*. Peterborough: Nature Conservancy Council.

Delderfield, E. R. 1974: *The Lynmouth Flood Disaster*. Exmouth: ERD Publications.

Department of the Environment (DoE) 1990: *Development on Unstable Land. PPG 14*. London: HMSO.

Department of the Environment (DoE) 1992a: *General Policy and Principles. PPG1*. London: HMSO.

Department of the Environment (DoE) 1992b: *Development Plans and Regional Planning Guidance. PPG12*. London: HMSO.

Department of the Environment (DoE) 1992c: *Coastal Planning. PPG 20*. London: HMSO.

Department of the Environment (DoE) 1992d: *Development and Flood Risk. Circular 30/92* (MAFF Circular FD1/92; Welsh Office Circular 68/92). HMSO, London.

Department of the Environment (DoE) 1993: *Emergency Financial Assistance to Local Authorities: Guidance Notes for Claims*. London: HMSO.

Department of the Environment (Doe) 1996: *Annex 1. Development on Unstable Land: Landslides and Planning. PPG 14*. London: HMSO.

Falconer, R. H. and Anderson, J. L. 1992: The February 1990 flood on the River Tay and subsequent implementation of a flood warning system. *IWEM Joint Meeting*, Perth.

Fowles, J. 1982: *A Short History of Lyme Regis*. Wimborne: Dovecote Press.

Hanwell, J. D. and Newson, M. D. 1970: The great storms of July 1968 on Mendip. *Wessex Cave Club Occasional Publication* 1(2).

Harrison, J. and Grant, P. 1976: *The Thames Transformed: London's River and its Waterfront*. London: André Deutsch.

High Point Rendel, in press: *The Investigation and Management of Soft Rock Cliffs*. London: Thomas Telford.

Hoskins, W. S. 1955: *The Making of the English Landscape*. London: Hodder and Stoughton.

Institution of Civil Engineers 1954: *Conference on the North Sea Floods of 31 January/1 February, 1953*. London: ICE.

Jensen, H. A. P. 1953: Tidal inundations past and present. *Weather*, 8, 85–9, 108–12.

Jones, D. K. C. and Lee, E. M. 1994: *Landsliding in Great Britain*. London: HMSO.

Kidson, C. 1953: The Exmoor storm and the Lynmouth floods. *Geography*, 38, 1–9.

Labadz, J. C., Butcher, D. P. and Potter, A. W. R. 1991: *Moorland Erosion in the Southern Pennines*, Part One. Research Monograph No. 1. Department of Geographical and Environmental Sciences, The Polytechnic of Huddersfield.

Lamb, H. H. 1991: *Historic Storms of the North Sea, British Isles and Northwest Europe*. Cambridge: Cambridge University Press.

Lee, E. M. 1993a: Landslides in Great Britain: investigation and management. *Structural Survey*, 11, 258–72.

Lee, E. M. 1993b: *Coastal Planning and Management: a Review*. London: HMSO.

Lee, E. M. 1995a: *The Occurrence and Significance of Erosion, Deposition and Flooding in Great Britain*. London: HMSO.

Lee, E. M. 1995b: *The Investigation and Management of Erosion, Deposition and Flooding in Great Britain*. London: HMSO.

Lee, E. M. 1995c: *Coastal Planning and Management: a Review of Earth Science Information Needs*. London: HMSO.

Lee, E. M. and Moore, R. 1991. *Coastal Landslip Potential: Ventnor, Isle of Wight*. Newport Pagnell: GSL Publications.

Lee, E. M. and Meadowcroft, I. C. in press: *The Cost of Flooding and Coastal Erosion in Great Britain*.

Lelliott, R. E. L. 1989: Evaluation of the Bournemouth defences. In *Coastal Management*. London: Thomas Telford, 263–77.

Marshall, W. A. L., 1952: The Lynmouth floods. *Weather*, 7, 338–42.

Ministry of Agriculture, Fisheries and Food/Welsh Office (MAFF) 1993a: *Strategy for Flood and Coastal Defence in England and Wales*. London: MAFF Publications.

Ministry of Agriculture, Fisheries and Food/Welsh Office/Welsh Office (MAFF) 1993b. *Project Appraisal Guidance Notes*. London: MAFF Publications.

Ministry of Agriculture, Fisheries and Food (MAFF) 1994: *Coast Protection Survey of England. Survey Report – Volume 1, Summary Report*. London: MAFF Publications.

Ministry of Agriculture, Fisheries and Food (MAFF) 1995. *Shoreline Management Plans: a Guide for Operating Authorities*. London: MAFF Publications.

Ministry of Agriculture, Fisheries and Food (MAFF) 1998: Memorandum submitted to the Agriculture Committee. In *Flood and Coastal Defence*, Vol. 2. London: The Stationery Office, 194–207.

Mirrlees, S. T. A., 1928: The Thames floods of January 7th. *Meteorological Magazine*, 63(2), 17–19.

Nairne, D. 1895: *Memorable Floods in the Highlands During the Nineteenth Century*. Northern Counties Printing and Publishing.

Newson, M. D. 1975: *Flooding and Flood Hazard in the UK*. Oxford: Oxford University Press.

Parker, D. J. 1988: Emergency service response and costs in British floods. *Disasters*, 12, 1–69.

Parker, R. 1978: *Men of Dunwich: the Story of a Vanished Town*. London: Holt, Rinehart and Winston.

Penning-Rowsell, E. C., Parker, D. J. and Harding, D. M. 1986: *Floods and Drainage*. London: George Allen & Unwin.

Pethick, J. 1996: Coastal slope development: temporal and spatial periodicity in the Holderness Cliff Recession. In M. G. Anderson and S. M. Brooks (eds), *Advances in Hillslope Processes*, Vol. 2. Chichester: Wiley, 897–917.

Pethick, J. and Burd, F. 1993: *Coastal Defence and the Environment*. London: MAFF Publications.

Purseglove, J. 1989: *Taming the Flood*. Oxford: Oxford University Press.

Robinson, D. A. and Blackman, J. D. 1990: Some costs and consequences of soil erosion and flooding around Brighton and Hove, autumn 1987. In J. Boardman, I. D. L. Foster and J. A. Dearing (eds), *Soil Erosion on Agricultural Land*. Chichester: Wiley, 369–82.

Robinson, D. N. 1969. Soil erosion by wind in Lincolnshire, March 1968. *East Midlands Geographer*, 4, 351–62.

Sear, D. A. and Newson, M. D. 1992: *Sediment and Gravel Transportation in Rivers Including the Use of Gravel Traps*. NRA project report. 232/1/T.

Steers, J. A. 1953: The east coast floods January 31–1 February 1953. *Geographical Journal*, 119, 280–98.

Taylor, S. 1994: Chichester's floods: a natural disaster? *Geographical Magazine*, May, 43–5.

Valentin, H. 1954: Der landverlust in Holderness, Ostengland von 1852 bis 1952. *Die Erde*, 6, 296–315.

Volker, M. 1953: La marée de tempête du 1er février 1953 et ses conséquences pour le Pays-Bas. *La Houille Blanche*, 2, 207–16.

Waverley Committee 1954: *Report of the Departmental Committee on Coastal Flooding*. London: HMSO.

Welsh Consumer Council 1992: *In Deep Water*. Cardiff: Welsh Consumer Council.

Willis Faber and Dumas Ltd 1996: *Research Report UK East Coast Flood Risk*. Ipswich: Willis Faber and Dumas Ltd.

Youdale, M. 1989: Wall of death. *Yorkshire Post*, February.

Chapter 10

Geomorphology for the Third Millennium

David L. Higgitt and E. Mark Lee

10.1 The Geomorphology of the Last Millennium: a Review

The picture of geomorphological activity in Britain that emerges from
the chapters of this book suggests that if it were possible to ignore the
land use changes and spread of development, the shape of the land
would have changed very little over the last 1000 years. Despite the
well-documented variability in climate and the increasing impact of
humans, the changes have only involved the minor modification of an
existing landscape inherited from the Quaternary and the Tertiary. At
a broad scale, this is a reflection of the stability of the landscape, with
negligible tectonic activity and an absence of prolonged periods of
extreme climatic conditions. However, at a more detailed level, changing patterns of sediment availability, together with variations in run-off
generation and sea-level rise, have led to locally significant adjustments
in form. The most pronounced changes have occurred along alluvial
channels, around the margins of upland areas and on coastlines dominated by shingle beaches, sand dune–beach systems and saltmarsh–
mudflat systems.

In reviewing a selection of the available literature for particular
process environments, the content of the volume demonstrates that
there has been a marked increase in attention paid to late Holocene
environmental change. At the outset, Brunsden (chapter 2) remarks
that the evidence for a clear geomorphological response to environmental change in the sediments of footslopes, valley floors and lakes is
disappointing, perhaps partly reflecting limitations in the resolution of
studies. Nevertheless, progress in untangling the chronology of late
Holocene process activity has been rapid, boosted by improvements in
the techniques available to decipher morphological and sedimentolog-

ical features. Rumsby (chapter 4) reviews the methods available to gather evidence for valley-floor activity over the last 1000 years. Technical developments open up new opportunities for compiling information that was hitherto unavailable. For example, the Global Positioning System (GPS) provides an efficient means for compiling mapping or survey data rapidly that can be used to assist geomorphological mapping (Higgitt and Warburton, 1999) or, as illustrated in chapter 4 (figure 4.1), to create detailed DEMs of channel and floor topography that would not be feasible with conventional surveying methods. Although novel techniques have added fresh perspectives to old problems, they are in themselves no substitute for the formulation of insightful hypotheses about the mechanisms and timing of geomorphological change.

A persistent theme throughout the book has been the extent to which geomorphological processes have responded to climatic variability and to human activities. Limitations in the precision of dating control have tended to result in a causal link between the observed morphological or sedimentological field evidence and the driving force being inferred from a coincidence in timing. In general terms, the confidence with which field evidence for enhanced process activity can be linked to a specific human-induced cause works best in small headwater catchments. Some examples are provided by Higgitt et al. (chapter 7) for several areas of upland Britain, notwithstanding recent findings that phases of alluvial fan formation in Scotland probably relate to a small number of individual storm events (Ballantyne and Whittington, 1999). Further downstream the anthropogenic signal is less clear and the valley-floor sequences have a tendency to reflect hydroclimatic variations. This reflects the influence of catchment scale, the importance of coupling between hillslope and channel transport and the relative sensitivity of slope and channel sediment sources to activation. Rumsby (chapter 4) notes that phases of marked channel incision in several upland valleys occurred in the eleventh and eighteenth centuries. The latter phase appears to correlate well with increased flood frequency. By contrast, few lowland rivers exhibit much change from this period, having become inhibited by extensive structural controls and imprisoned in cohesive banks. Thus the spatial variability of the sensitivity of fluvial systems to change is a marked feature of the last millennium, a theme portrayed through a series of short cases studies by Hooke (chapter 5).

On hillslopes, Jones (chapter 3) finds some support from the documentary record of landslides for some association of enhanced activity and the timing of the Little Ice Age, but the evidence is ambiguous. In the uplands, where documentary evidence for slope instability is scarce and the scope for reconstructing chronology limited, there are a number of lines of evidence suggesting that accelerated erosion occurred from the mid-nineteenth century onwards, postdating the coldest phases of the Little Ice Age, but closely associated with increasing numbers of sheep. In lowland rural environments, analysis of sediment yields also suggests that sediment transfer has accelerated over the last century, which can be linked to both land use change and variability in the seasonality of cyclonic weather (Foster, chapter 8).

Whereas landform change has been somewhat limited in extent and subtle in expression across many land surfaces, the coastal geomorphology of Britain over the last 1000 years has been characterized by dynamic change. Against a model of continued adjustment to interglacial conditions that is presented by Lee (chapter 6), there is some evidence for coastal processes responding to climate change within the last millennium. A slight increase in sea level in the early part of the last millennium (coincident with the Medieval Warm Period) initiated accelerated erosion of soft cliffs and the inland expansion of saltmarshes in estuaries. This might present an analogue for the present day, but for the extent to which human activities have disrupted the sediment transfer dynamics of coasts and estuaries, with major implications for coastal protection. Inland, Jones (chapter 3) calculates that the deliberate removal of material by human agency far exceeds transport by natural processes. The emergence and acceleration of human impact on the British landscape during the last 1000 years is an important theme which is developed below.

10.2   Human Impact

The character of many areas has been modified, to varying degrees, by the human occupancy of the landscape and land management practices. The extent and significance of these modifications has accelerated through the millennium as population has grown, urban areas have expanded into the countryside and land use has become more intense. Amongst the more significant influences over the last century are those associated with changing socio-economic factors, including the following:

1   *Agricultural practice.* The general postwar intensification of agricultural production has been accompanied by removal of hedgerows, use of heavy vehicles, monoculture and changing land management practice. All are believed to have contributed to an increase in soil erosion. There is general agreement that the reported increase in

significant lowland erosion events is due to the adoption of winter cereals and the consequent expansion of the area left bare in autumn and winter (Boardman and Robinson, 1985; Evans and Cook, 1986; Speirs and Frost, 1987). Indeed, the area sown to winter cereals has increased by a factor of over three since 1969.

2   *Forestry.* After the Second World War, timber production was seen as of strategic importance, with the result that large areas of upland Britain have been afforested in the last 40 years or so. Between 1945 and 1983, 700 000 ha was planted by the Forestry Commission, representing the single largest land use change in Britain over this period (Acreman, 1985). Around 60% of this land has been ploughed prior to planting (Taylor, 1970). It is believed that this has led to temporary increases in hillslope erosion and supply of sediment to stream channels, Research suggests increased stream sediment loads of up to 1600% in forested areas compared with non-forested areas (Soutar, 1989). The increased sediment loads are suspected to have had an impact on salmon and trout rivers. For example, Drakeford (1979) correlated the reduction in catches of salmonoid fish in the River Fleet with the expansion of forestry in the catchment, noting a 90% decline in sea trout catches between 1960 and 1978.

It is possible that upland forestry has led to changes in flood behaviour further down a catchment. The extensive drainage operations necessary for planting in peaty areas can have immediate effects on slope hydrology by increasing run-off and flow in the upland streams, and reducing the time to reach peak flow. This effect is enhanced by the practice of ploughing downslope, creating numerous artificial channels. Robinson (1981) demonstrated that in the first seven years after ploughing a north Pennine catchment for forestry, there were higher flood flows and the time to peak flow was halved.

3   *Water supply.* The supply of cheap and clean water for industry and lowland centres of population from upland areas has transformed the behaviour of many rivers. Reservoir storage is essential to ensure a regular supply and around 450 dams have been constructed in Britain (Petts, 1988). Many water supply reservoirs were built upstream of industrial towns. Petts (1989) reports that disputes over water usage led to the need for compensation reservoirs downstream or in tributary valleys to maintain compensation flows to mills. Major reservoirs were constructed in the nineteenth century in the upper Severn and Wye valleys to supply Liverpool and Birmingham, respectively (Sheail, 1984). These structures have also helped regulate flood flows. However, there have also been down-

stream implications for river channel form (Petts, 1979). Generally, dams decrease peak flows downstream and impound practically all sediment. Channels tend to reduce in size downstream, but also tend to incise because of the lack of sediment.

4   *Mining operations.* Subsidence associated with the collapse of abandoned mine workings has been widespread throughout the coalfield areas and other mining centres (e.g. the salt mine areas of Cheshire). In South Wales a number of major landslides coincided with the industrial development of the area and are believed to have been associated with the collapse of shallow mine workings. The disposal of mine waste has also led to slope instability problems, notably the flowslides on colliery spoil tips in South Wales (Jones and Lee, 1994): the Aberfan disaster of October 1966 has been the most tragic example, with 144 fatalities.

Former mine workings in upland areas frequently disposed of spoil and waste in mounds adjacent to river channels or directly into the stream itself. The resulting increase in sediment load can have an impact on its stability, as shown in several valleys in northeastern England and in central Wales (Lewin et al., 1983; Macklin and Lewin, 1989; Howard et al., 2000). Generally, increased sediment lead to increased instability, tendency for braiding, aggradation and channel switching. The reduction in sediment loads following mine closure can reverse some of the channel changes, but adjustment will continue long after the cessation of mining (Macklin and Lewin, 1997).

5   *Development.* The increasing demand for housing and employment opportunities by a growing population has led to an increased utilization of floodplains and clifftop locations. Frequently, these are viewed as desirable settings for expensive housing. Development has been accompanied by: an increase in the amount of impermeable surfaces within a catchment, increasing and accelerating run-off; reduction in floodplain storage following flood defence works; changes in flood behaviour following the construction of sewage and stormwater drainage systems; and uncontrolled surface water discharge into slopes, leading to increased erosion problems.

Urbanization can lead to channel modifications downstream following an increase in the size of peak flows and a decrease in sediment supply (Knight, 1979; Hollis and Luckett, 1976). Amongst the most common changes are bed and bank erosion, loss of riverbank trees and undermining of structures. In a review of channel changes following the development of Cumbernauld New Town in Strathclyde, Roberts (1989) noted that some streams had enlarged through the urban area. Locally, there were instances of extensive erosion and bank collapse; up to 10 m of vertical incision through glacial till was observed in gulleys draining industrial areas. Some channels developed gravel bars and became more braided.

Modern suburbs tend to have large-diameter drains, allowing rapid run-off direct to the nearest watercourse. This can lead to an increase in flows in the channel and a quicker response to rainfall events. The size of change is related to the extent of development and the return period of a particular event; small events are affected most. A one year return period event can be enhanced ten times for 40% urbanization, whereas the two year event is only doubled or trebled (Hollis, 1975). At both Stevenage and Skelmersdale, for example, Knight (1979) found that the mean annual flood was about 2.5 times its former size after urbanization.

Expansion of development on to floodplains reduces the storage capacity of a river system. Floodplain storage is an important mechanism by which a flood wave can be dampened as it travels down a reach, reducing the peak of the flood but extending its duration (i.e. the flood is spread out over a longer period and, hence, is less severe). Flooding of rural areas can be important for the relief of flooding in urban areas downstream, although it can be detrimental to agricultural interests.

Development can have a significant effect on slope stability. One of the most serious effects is often the artificial recharge of the groundwater table and, hence, increased porewater pressures. At Ventnor, Isle of Wight, for example, it has been shown that uncontrolled discharge of surface water through soakaways and highway drains may have contributed to raising the groundwater table to a level at which heavy winter storms may trigger movement (Lee and Moore, 1991). In addition, progressive deterioration and leakage of swimming pools and services such as foul sewers, storm sewers, water mains and service pipes can all contribute to instability problems. The example of Luccombe, Isle of Wight, illustrates this point, whereby leakage from septic tanks and water supply pipes was identified as a cause of the 1987–8 ground movements (Lee and Moore, 1989).

6   *Land reclamation.* Extensive land reclamation has taken place since Roman times in many estuaries, often involving the enclosure of saltmarshes to control tidal flooding and improve grazing conditions. Over time, urban and industrial uses have replaced agriculture in many estuaries. It is estimated that over 100 000 ha of intertidal land has been reclaimed (Davidson et al., 1991). The impacts on estuary processes have included the modification of the tidal prism, with resultant changes in the pattern of saltmarsh/

mudflat erosion and accretion. Tidal delta structures have collapsed in response to the changing tidal prisms, with consequent effects on the open coastline. For example, the dramatic erosion and loss of Dunwich, Suffolk, may have been accelerated by reclamation works in the adjacent estuaries.

7   *River channelization and flood defences.* These have often led to modifications to the patterns of erosion and deposition along a stretch of river, leading to river channel migration. Flood defence embankments to protect vulnerable communities are designed to prevent floodplains acting as stores and can, therefore, create problems in downstream reaches, where the loss of flood wave attenuation can lead to more severe flooding than may have previously been the case. Flood embankments can prevent the over-bank deposition of suspended sediments, reducing floodplain accretion rates and leading to sedimentation problems downstream and constraining channel metamorphosis.

In estuaries, the progressive constriction of the channel through the construction of embankments can lead to a reduction in the storage capacity of the estuarine lowlands and marshes. Horner (1978), for example, has demonstrated that high tides on the Thames at London Bridge have risen dramatically over the last 150 years: this can only be partly explained by rising sea levels and settlement in the London area caused by groundwater abstraction. The reduction in storage can act to pass the peak of the flood tide upstream, increasing the potential flood problems above the reaches that have been defended; that is the opposite of the effects of flood defence on rivers.

8   *Coastal defences.* These have frequently resulted in a disruption in the supply and transport of sediment around the coast. This, in places, has led to increased coastal erosion or flood risk, especially where natural defences such as sand dunes or beaches are deprived of a regular supply of sediment. Coastal instability can also be exacerbated by the disruption of sediment transport, as occurred at Folkestone Warren following harbour construction (Hutchinson, 1969; Hutchinson et al., 1980). The most recent phase of landslide activity was probably a consequence of the expansion of the Folke-stone Harbour facilities over the years 1810–1905. In particular, the extension of the main pier resulted in the disruption of wave and current induced littoral drift of sand and shingle eastwards through the area, leading to the build up of trapped material west of the pier and beach shrinkage through undernourishment to the east, at the foot of the Warren. Reduced beach volume would have led to increased wave erosion of the toes of pre-existing rotated

blocks, which slid seawards, thereby removing support from the base of the cliffs.

The impacts of humans can be evaluated in terms of their modification of the cascade of sediment from hillslopes through a river network to the coast. The modifications within catchments include the following:

1 *Changes in the pattern of sediment supply.* Accelerated erosion, especially in recently afforested upland areas, lowland areas under winter cereals and along some alluvial river channels needs to be set against the 'sterilization' of extensive potential sediment supply areas by built development and river channelization.

2 *Changes in the delivery of sediment from hillsides to river channels.* Hedgerows, ditches and urban infrastructure all act as barriers to sediment transport. Measurements of soil erosion in southern England suggest that between 5% and 15% of arable land may be affected by erosion in any one year. Although between 25% and 90% of the eroded material may be transported beyond the boundaries of the eroding field (Quine and Walling, 1991), much of this material will be deposited before it reaches a watercourse. The artificial barriers have been more effective in trapping coarse sediment, leading to a change in the relative proportions of coarse and fine sediment delivery to rivers.

3 *Changes in sediment transport along rivers.* On many rivers and streams the transport of coarse sediment has been severely restricted by artificial structures, such as weirs, reservoirs and discharge control structures, which act as sand and gravel traps. The removal of coarse material from river beds, either as part of flood defence maintenance (Sear and Newson, 1992) or for supply of aggregates for construction (Fleming, 1970) also reduces the actual downstream transport.

4 *Changes in the delivery of sediment from river channels to floodplains.* The construction of flood embankments along large sections of many rivers has reduced the frequency of floodplain inundation and, hence, the deposition of fine sediments. The absence of overbank transmission of fine sediment leads to higher loads within the main channel, with the net result of increased within-channel siltation.

5 *Changes in the delivery of sediment from rivers to the open coast.* Modifications of river mouths and estuaries have ranged from land reclamation, capital and maintenance dredging of navigation channels, port and harbour development and mud-digging to the con-

struction of tidal barrages. All of these activities will have had an influence on the patterns of deposition and erosion of fine sediment, and the transport of coarse sediment bedload. It is likely that many rivers now have little or no coarse sediment connection to the coast and only deliver fine sediments.

On the coast, the impacts of human activity have included the following:

1 *Reduction in sediment inputs from cliff recession.* Over the last 100 years or so, some 860 km of coastal protection works have been constructed to prevent cliff recession. Although difficult to quantify, sediment inputs could have declined by as much as 50% over this period. For example, on the North Yorkshire coast between Whitby and Sandsend it has been estimated that current sediment inputs from the eroding glacial till cliffs may be around 35% of the inputs prior to the onset of coast protection works in the late nineteenth and early 20th centuries. On a national level, this decline has probably been a factor in the degradation of beaches around many parts of the coastline, As Clayton (1980) wrote about the Anglian coastline:

> Comparison of areas lost and gained by coastal erosion in the early years of this century with recent years, shows that the building of coastal defences along part of the coast was effective in reducing net land loss. At that time selection of relatively easy sites coupled with the continuing drift of sand from feeder bluffs showed the advantage in terms of net area to be gained by accepting the erosion of high cliffs and consolidating the progradation of neighbouring low-lying areas. Public pressure, willingly accepted by engineers, had brought extension of defences to 60% of the coast, but no improvement in the erosional balance. Indeed, the successful stabilisation of some of the lower cliffs has removed local sand sources, so that the longer transport paths from the major feeder bluffs are now dominant. Our work suggests that . . . the removal of these inputs of sand to the system would initiate a decline in beach volumes that could build up to catastrophic proportions.

2 *Disruption of longshore sediment transport.* Groynes, harbour break-waters and other shoreline structures have had a significant impact on coastal processes. For example, apparent increases in cliff erosion at Fairlight, East Sussex during the 1980s are believed to have been related to a diminishing supply of shingle arriving at the base of the cliff to provide natural protection against wave attack (Pen-

ning-Rowsell et al., 1992). This diminishing supply is attributed to the construction of harbour and coast protection works to the west, including the defences of Hastings, which were constructed between 1830 and 1930. Elsewhere, large beaches have built up behind these artificial barriers, as illustrated by the growth of Monmouth beach since the connection of The Cobb, Lyme Regis to the shoreline in the eighteenth century.

3   *Disruption of cross-shore sediment transport.* Beaches and sand dunes form part of an integrated sediment exchange system. Sand may be deposited on the upper beach during low wave energy events, providing a source of sediment for aeolian transport and foredune growth. During high-energy events, the beach will flatten and the foredunes will erode. This provides an additional source of sand that is transported seaward and deposited on the lower beach. However, in many areas the response to foredune erosion has been to construct sea walls or rock revetments, thus severing the linkage between the dune and beach, to the detriment of both.

4   *Decline in the volume of shoreline sediment stores.* Beach mining for sand and gravel has had a significant impact and many beach and dune areas. For example, gravel was probably extracted from Chesil Beach for at least 700 years, with an estimated 1.1 M tonnes (0.7 Mm³) removed between the mid-1930s and 1977. Since 1820 it is possible that the volume of Chesil Beach has declined by around 930 000 m³ of shingle; these losses could represent between 1.5% and 5% of the estimated beach volume (Carr and Blackley, 1974). In Scotland, large-scale sand extraction is common on the dune and beach complexes of the Orkney and Shetland Islands (e.g. the Bay of Quendale) and, to a lesser extent, around Brodick and Girvan on the Clyde coast. It has been estimated that sand extraction has had an adverse effect on 16% of all beach complexes in Scotland (Ritchie and Mather, 1954).

5   *Decline in offshore sediment stores and sinks.* The marine aggregates industry supplies about 15% of the UK demand for sand and gravel (Lee, 1993). Much of the 20M tonnes landed each year is extracted off the East and South Coasts of England, with smaller amounts off the west coast of Wales (Nunny and Chillingworth, 1986). Seabed material is in increasing demand for beach replenishment schemes.

The net effect of these modifications has been fragmentation of sediment transport pathways and cells. This is most obvious on the coast, where significant morphological adjustments have taken place because of the resulting changes in sediment availability.

10.3   Geomorphological Challenges

What opportunities and challenges does the analysis of millennial-scale geomorphological change provide for the future direction of the subject? There are implications for both the academic development of the subject and its application in environmental management. A subjective list of topics is highlighted briefly below.

10.3.1   Geomorphology and climate change

The landforms of Britain exhibit considerable variability in their apparent sensitivity to environmental change. The widespread floods in 2000 have brought issues of future climate change to the fore, amid public concerns that structural defence from natural processes should be enhanced. Several of the chapters have illustrated how structural interventions in geomorphological systems have constrained geomorphological responses but had knock-on consequences for sediment transfer and associated ecological processes. The development of a management framework for dealing with predicted climate change presents a dominant challenge. Although climate change has attained public attention, attempts to predict regional change is fraught with difficulty. In particular, Britain is sensitive to changes in the location of rain-bearing depression tracks, whose future alteration is difficult to ascertain. It is likely that the most significant changes will involve increased cyclonicity in the north and west, with milder and wetter winters. Southern Britain is likely to experience wetter winters but with a marked decrease in summer run-off and an increased potential for summer storms. Geomorphologically, the consequences of climate change go far beyond the hydroclimatic variation that may result, because the land use changes that will accompany a warming trend will have implications for sediment transfer as the potential for soil erosion increases (Boardman and Favis-Mortlock, 1993). Having established a network of flood protection schemes during the twentieth century to deal with prevailing flood frequencies, river managers of the future are likely to be facing the challenge of controlling rivers with more significant erosion and sedimentation problems (Newson and Lewin, 1991). At the coast, the geomorphological response to rising sea levels is likely to be constrained by existing lines of defence, raising issues for engineering strategies.

10.3.2   Geomorphology and history

The persistence of human impact as a factor in explaining geomorpho-logical process activity during the last millennium indicates the intricate links between the operation of natural processes and the history of landscape development. It was stated at the outset that there has been a tendency for the study of the historical development of scenery to be conducted separately from the study of environmental change. Indeed, in chapter 5, Hooke remarks that it is only within recent decades that geomorphologists have turned their attention to looking for evidence of change over historical periods. By contrast, historians have had a long involvement in describing and explaining the evolution of scenery. At a time when there would appear to be considerable opportunities for collaborative work (as exemplified by Hooke and Kain, 1982), it is unfortunate that historical geographers appear to have taken a cultural turn away from field-based research. Recent years have witnessed a growing collaboration between geomorphologists and archaeologists (e.g. Needham and Macklin, 1992; Pollard, 1999; Dincauze, 2000). Geomorphological interpretation of alluvial sequences provides archae-ologists with information about likely hydrological and environmental conditions at the time of occupation, while the artefacts provide additional clues for chronology. As part of the quest for geomorpholo-gists to improve the resolution and detail of historical reconstruction, a developing dialogue with historical geographers, historians and indus-trial archaeologists offers the prospect of improved understanding of the combined effect of anthropogenic and geomorphologiocal changes and of the reactions, if any, of human settlement to environmental conditions.

10.3.3   Geomorphology and engineering

Although much of the landscape of Britain has been shown to be ancient, benign, stable and insensitive to change, the operation of geomorphological processes continue to present a range of management issues. Brunsden (chapter 2) warns that the high energy absorption capacity of British landforms masks the ability for slow and progressive forces to reactivate apparently stable systems. Many of the inland landslides that have impacted upon land development schemes result from human disturbance of relict forms. Analysis of the historical record of natural hazards reveal an extensive catalogue of geomorpho-logical events that have caused considerable financial and human costs to society. The nature of these hazards have been reviewed by Lee (chapter 9). The response of society to such hazard – whether as individuals, enterprises or through the machinery of the state – has often involved a considerable amount of engineering. The corroboration between geomorphology as a means of establishing the processes that lead to instability or the magnitude and frequency characteristics of specific hazards and engineering as a means of managing the risk would appear obvious, but it is a union that has only recently been developed. The evidence produced throughout the volume for the way in which process systems have responded over the 1000 year timescale is directly relevant to engineering design where the design event might have a 500 year or 1000 year recurrence interval.

The growing profile of geomorphological analysis in an engineering context (Thorne et al., 1997; Thorne, 1998) is mirrored by the emerging concepts of environmental (or 'soft') engineering. This is perhaps best exemplified by river restoration techniques which seek to provide design that is sympathetic to natural form and function (Brookes, 1995; Kondolf, 1996). As shown by Brunsden (chapter 2, figure 2.7) and developed by Hooke (chapter 5), most British lowland rivers have been heavily modified by centuries of human involvement. Ultimately restoration offers the complete structural and functional return to a pre-disturbance state (Cairns, 1991), but in practice there are many obstacles to full restoration and most schemes are limited to enhancing certain aspects of the structural or functional attributes. In Britain, river restoration has been piloted through the River Restoration Project (now known as the River Restoration Centre, RRC), commencing in 1994 and aiming to establish state-of-the-art demonstration sites. With the support of EU funding, two demonstration sites have been developed on the urban River Skerne in Darlington and on the rural River Cole in Oxfordshire. By 1999 it was estimated that some 350 river regeneration schemes were in place across Britain (River Restoration Centre, 1999). River restoration is an example of how hydrological and geomorphological analysis has taken a higher profile in practical river management and encouraged greater partnerships between geomorphologists, engineers, ecologists and the relevant local communities and funding agencies.

The dialogue between geomorphology and ecology which is fundamental in river restoration has grown steadily in the last decade (Gardiner, 1991; Boon et al., 1992), bolstered by the recognition that the transfer of water, sediment and nutrients by geomorphological processes has a profound influence on the creation, maintenance and quality of habitat and species diversity. In the context of a river restoration engineering scheme, the links between fluvial geomorphology and habitat are apparent. Straightened channels usually have over-steepened banks and bank re-profiling creates habitat and reduces shading that encourages species diversity; pool and riffle construction and the reinstatement of varied substrate material promotes fish spawning and creates niches for different stages of the fish life-cycle; and reinstatement of meandering improves habitat, plant biomass and invertebrate numbers, which in turn support the fish community (Brookes and Shields, 1996).

Similar examples for coastal environments are developed by Lee (chapter 6), where the prevalence of embankment construction has inhibited sediment transfer processes and resulted in dramatic salt-marsh erosion and the loss of intertidal habitats in some estuaries. Although structural control dominated coastal management throughout the twentieth century, the last decade has seen significant changes in attitudes towards the use of engineering, manifest in the recognition of a need for strategic planning of flood and coastal defence issues through Shoreline Management Plans (MAFF, 1995) and the advocacy of soft engineering approaches on the coastline, including the notion of managed retreat (Pethick and Burd, 1993).

10.3.4   Geomorphology and landscape conservation

Following directly from the growing interest in reconstructing artificial landforms that are geomorphologically sympathetic as well as functional is the role of management in preserving and enhancing natural landscape features. The preservation of sites of biological or archaeological interest has a long pedigree, but recognition of sites that are important for geological or geomorphological features is more recent. Emphasis has not only been placed upon good examples of geological or landform features but also on preserving type sites which led to the development of ideas in Earth science. Challenges in geomorphological conservation arise from the recognition that many features of interest are relict forms from past, particularly Late Glacial, conditions, while others evolved during the transition to interglacial conditions. With regard to montane landforms, Higgitt et al. (chapter 7) note that management may be necessary to protect relict forms from disturbance by contemporary geomorphological processes. Gordon et al. (1998) evaluate how the spectacular assemblage of geomorphological features in the Cairngorm Mountains is under threat from a range of human activities, for which a variety of statutory and voluntary measures are in place to provide environmental protection.

The conservation imperative of the landforms is enhanced further by recognition of the strong association between landform assemblage and habitat zones. Geomorphology is challenged with the development of an inclusive concept of nature conservation that is concerned not only with plant and animal communities but with the physical landscape in which those habitats are developed and maintained. While measures can be taken to impose environmental protection at specific sites, the impacts of climate change and pollution generally lie beyond the control of local management. Foster (chapter 8) highlights the contribution that geomorphological studies have made to understanding the transfer of sediment-associated nutrients through lowland catchments. A further challenge to geomorphology is to improve understanding of transfer dynamics as a means for encouraging land management practices that reduce downstream impacts on water quality. Some practical success has been made in designing buffer strips to control nitrate pollution (Haycock and Burt, 1993), but the development of models of particulate nutrient and contaminant transfer, particularly at catchment scale, hinges upon many unanswered questions.

10.3.5   Geomorphology and public opinion

Despite the success of projects such as the River Restoration Programme in generating co-operation between geomorphologists, ecologists and engineers, the expense associated with restoration programmes is such that it is unlikely that many schemes will be initiated. Meanwhile concerns that flooding is likely to be a more frequent feature of a warming twenty-first century develop political pressure for more structural control. Despite a detectable shift in public policy that Lee (chapter 9) characterizes as a transition from 'nature commanded' towards 'living with processes', it would seem that public desire for defence from natural hazards will accentuate the structural constraint on river channels and coasts. There is competition between delivering a high standard of protection and permitting fluvial or coastal processes to operate naturally and generate physical and biological diversity. Hooke (chapter 5) laments that geomorphologists of the future will inherit a sterile and artificial fluvial landscape, where continued emphasis on structural control will have grossly inhibited the natural functions of erosion and sedimentation in channels. Such a conclusion would be a depressing note on which to end this volume. Rather, the renewed interest in managing a changing environment that has arisen from the storms and floods in the first year of the third millennium provides a context for an enhanced contribution of geomorphology. It can be argued that, unlike ecology, history and archaeology, which have successfully developed a popular profile through television features, there is relatively little public consciousness of the nature and practice of geomorphology. The project from which this book arises was undertaken in the belief that the story of the impact of geomorphological process activity on the British landscape over the last 1000 years, its conjunction with the historical development of the countryside and its implications for the challenges of future environmental management is one that has been worth telling. The ultimate challenge is to raise awareness of the function of geomorphological processes in landscape and environmental management in the minds of policy-makers and the general public.

REFERENCES

Acreman, M. C. 1985: The effects of afforestation on the flood hydrology of the Upper Ettrick Valley. *Scottish Forestry*, 39, 89–99.
Ballantyne, C. K. and Whittington, G. 1999: Late Holocene floodplain incision and alluvial fan formation in the central Grampian Highlands, Scotland: chronology, environment and implications. *Journal of Quaternary Science*, 14, 651–71.
Boardman, J. and Favis-Mortlock, D. T. 1993: Climate change and soil erosion in Britain. *Geographical Journal*, 159, 179–83.
Boardman, J. and Robinson, D. A. 1985: Soil erosion, climatic vagary and agricultural change on the Downs around Lewes and Brighton, Autumn 1982. *Applied Geography*, 5, 243–58.
Boon, P. J., Petts, G. E. and Calow, P. (eds) 1992: *River Conservation and Management*. Chichester: Wiley.
Brookes, A. 1995: River channel restoration: theory and practice. In A. Gurnell and G. E. Petts (eds), *Changing River Channels*. Chichester: Wiley, 370–87.
Brookes, A. and Shields, F. D., Jr. 1996: *River Channel Restoration – Guiding Principles for Sustainable Projects*. Chichester: Wiley.
Cairns, J. 1991: The status of the theoretical and applied science of restoration ecology. *The Environment Professional*, 13, 186–94.
Carr, A. P. and Blackley, M. W. L. 1974: Ideas on the origin and development of Chesil Beach, Dorset. *Proceedings of the Dorset Natural History and Archaeological Society*, 95, 9–17.
Clayton, K. M. 1980: Coastal protection along the East Anglian coast. *Zeitschrift für Geomophologie*, 34, 165–72.
Davidson, N. C., d'A Laffolely, D., Doody, J. P., Way, L. S., Gordon, J., Key, R., Drake, C. M., Pienkowski, M. W., Mitchell, R. and Duff, K. L. 1991: *Nature Conservation and Estuaries in Great Britain*. Peterborough: Nature Conservancy Council.
Dincauze, D. F. 2000: *Environmental Archaeology: Principles and Practice*. Cambridge: Cambridge University Press.

Drakeford, T. 1979: *Report of Survey of the Afforested Spawning Grounds of the Fleet Catchment.* Forestry Commission.

Evans, R. and Cook, S., 1986: Soil erosion in Britain. *SEESOIL*, 3, 28–58.

Fleming, G. 1970: Sediment balance of the Clyde estuary. *Proceedings of the American Society of Civil Engineers, Hydraulics Division*, 96, 2219–30.

Gardiner, J. L. 1991: *River Projects and Conservation: a Manual for Holistic Appraisal.* Chichester: Wiley.

Gordon, J. E., Thompson, D. B. A., Haynes, V. M., Brazier, V. and Macdonald, R. 1998: Environmental sensitivity and conservation management in the Cairngorm Mountains, Scotland. *Ambio*, 27, 335–44.

Haycock, N. E. and Burt, T. P. 1993: Role of floodplain sediments in reducing the nitrate concentration of subsurface runoff: a case study in the Cotswolds, UK. *Hydrological Processes*, 7, 287–95.

Higgitt, D. L. and Warburton, J. 1999: Applications of differential GPS in upland fluvial geomorphology. *Geomorphology*, 29, 121–34.

Hollis, G. E. 1975: The effect of urbanisation on floods of different recurrence intervals. *Water Resources Research*, 11, 431–5.

Hollis, G. E. and Luckett, J. K. 1976: The response of natural streams to urbanisation: two case studies from south east England. *Journal of Hydrology*, 30, 351–63.

Hooke, J. M. and Kain, R. J. P. 1982: *Historical Change in the Physical Environment: a Guide to Sources and Techniques.* Sevenoaks: Butterworths.

Horner, R. W. 1978: The Thames Barrier project. *Geographical Journal*, 154, 242–53.

Howard, A. J., Macklin, M. G., Black, S. and Hudson-Edwards, K. 2000: Holocene river development and environmental change in Upper Wharfedale, Yorkshire Dales, England. *Journal of Quaternary Science*, 15, 239–52.

Hutchinson, J. N. 1969: A reconsideration of the coastal landslides at Folkestone Warren, Kent. *Geotechnique*, 19, 6–38.

Hutchinson, J. N., Bromhead, E. N. and Lupini, J. F. 1980: Additional observations on the Folkestone Warren landslides. *Quarterly Journal of Engineering Geology*, 13, 1–31.

Jones, D. K. C. and Lee, E. M. 1994: *Landsliding in Great Britain: a Review.* London: HMSO.

Knight, C. R. 1979: Urbanisation and natural stream channel morphology: the case of two English new towns. In G. E. Hollis (ed.), *Man's Impact of the Hydrological Cycle in the United Kingdom.* Norwich: Geobooks, 181–98.

Kondolf, G. M. 1996: A cross section of stream channel restoration. *Journal of Soil and Water Conservation*, 51, 119–25.

Lee, E. M. 1993: *Coastal Planning and Management: a Review.* London: HMSO.

Lee, E. M. and Moore, R. 1989: *Landsliding in and Around Luccombe Village, Isle of Wight.* London: HMSO.

Lee, E. M. and Moore, R. 1991: *Coastal Landslip Potential Assessment: Isle of Wight Undercliff, Ventnor.* Geomorphological Services Ltd.

Lewin, J., Bradley, S. B. and Macklin, M. G. 1983: Historical valley alluviation in mid-Wales. *Geological Journal*, 18, 331–50.

Macklin, M. G. and Lewin, J. 1989: Sediment transfer and transformation of an alluvial valley floor: the River South Tyne, Northumbria, UK. *Earth Science Processes and Landforms*, 14, 233–46.

Macklin, M. G. and Lewin, J. 1997: Channel, floodplain and drainage basin response to environmental change. In C. R. Thorne, R. D. Hey and M. D. Newson (eds), *Applied Fluvial Geomorphology for River Engineering and Management*. Chichester: Wiley, 15–45.

Ministry of Agriculture, Fisheries and Food (MAFF) 1995: *Shoreline Management Plans: a Guide for Operating Authorities*. London: MAFF Publications.

Needham, S. and Macklin, M. G. (eds) 1992: *Alluvial Archaeology in Britain*. Oxbow Monograph 27. Oxford: Oxbow Press.

Newson, M. and Lewin, J. 1991: Climatic change, river flow extremes and fluvial erosion. *Progress in Physical Geography*, 15, 1–17.

Nunny, R. S. and Chillingworth, P. C. H. 1986: *Marine Dredging for Sand and Gravel*. London: HMSO.

Penning-Rowsell, E. C., Green, C. H., Thompson, P. M., Coker, A. M., Tunstall, S. M., Richards, C. and Parker, D. J. 1992: *The Economics of Coastal Management: a Manual of Benefits Assessment Techniques*. London: Belhaven Press.

Pethick, J. and Burd, F. 1993: *Coastal Defence and the Environment*. London: MAFF Publications.

Petts, G. E. 1979: Complex response of river channel morphology subsequent to reservoir construction. *Progress in Physical Geography*, 3, 329–62.

Petts, G. E. 1988: Regulated rivers in the UK. *Regulated Rivers*, 2, 201–20.

Petts, G. E. 1989: Historical analysis of fluvial hydrosystems. In G. E. Petts, H. Moller and A. L. Roux (eds), *Historical Change of Large Alluvial Rivers: Western Europe*. Chichester: Wiley, 1–18.

Pollard, A. M. (ed.) 1999: *Geoarchaeology: Exploration, Environments, Resources*. Special Publication 165. London: Geological Society.

Quine, T. A. and Walling, D. E. 1991: Rates of soil erosion on arable fields in Britain: quantitative data from caesium-137 measurements. *Soil Use and Management*, 7, 169–76.

Ritchie, W. and Mather, A. S. 1985: *The Beaches of Scotland*. Edinburgh: Countryside Commission for Scotland.

River Restoration Centre 1999: *Manual of River Restoration Techniques*, 1st edn. Bedford: Arca Press.

Roberts, C. R. 1989: Flood frequency and urban-induced channel change: some British examples. In P. Carling and K. Bevan (eds), *Floods: Hydrological, Sedimentological and Geomorphological Implications*. Chichester: Wiley, 57–82.

Robinson, M. 1981: *The Effects of Pre-afforestation Drainage upon the Streamflow and Water Quality of a Small Upland Catchment*. Wallingford: Institute of Hydrology.

Sear, D. A. and Newson, M. D. 1992: *Sediment and Gravel Transportation in Rivers Including the Use of Gravel Traps*. NRA project report 232/1/T.

Sheail, J. 1984: Constraints on water-resource development in England and Wales: the concept and management of compensation flows. *Journal of Environmental Management*, 19, 351–61.

Soutar, R. G. 1989: Afforestation and sediment yields in British fresh waters. *Soil Use and Management* 5, 82–6.

Speirs, R. B. and Frost, C. A. 1987: Soil water erosion on arable land in the United Kingdom. *Research and Development in Agriculture*, 4, 1–11.

Taylor, G. G. M. 1970: *Ploughing Practice in the Forestry Commission*. Forest Record No. 73. London: HMSO.

Thorne, C. R. 1998: *Stream Reconnaissance Handbook*. Chichester: Wiley.

Thorne, C. R., Hey, R. and Newson, M. D. (eds) 1997: *Applied Fluvial Geomorphology for River Engineering and Management*. Chichester: Wiley.

Geographical Index

Page numbers in *italics* refer to figures or tables.

Subject Index

Page numbers in *italics* refer to figures or tables.

Lightning Source UK Ltd.
Milton Keynes UK
UKOW041103120713

213692UK00001B/58/P